500

ELECTRONIC IC CIRCUITS WITH PRACTICAL APPLICATIONS

500

ELECTRONIC IC CIRCUITS WITH PRACTICAL APPLICATIONS

James A. Whitson

TAB TAB BOOKS
Blue Ridge Summit, PA

FIRST EDITION
FOURTH PRINTING

© 1987 by TAB BOOKS
TAB BOOKS is a division of McGraw-Hill, Inc.

Library of Congress Cataloging-in-Publication Data

Whitson, James A.
500 electronic IC circuits, with practical applications / by James
A. Whitson
p. cm.
Includes index.
ISBN 0-8306-7920-0   ISBN 0-8306-2920-3 (pbk.)
1. Integrated circuits—Handbooks, manuals, etc. I. Title.
II. Title: Five hundred electronic circuits, with practical
applications.
TK7874.W485   1987   87-26232
621.381′73—dc 19   CIP

TAB BOOKS offers software for
sale. For information and a catalog,
please contact TAB Software Department,
Blue Ridge Summit, PA 17294-0850.

Questions regarding the content of this book should be addressed to:

Reader Inquiry Branch
TAB BOOKS
Blue Ridge Summit, PA 17294-0214

Contents

Acknowledgments

I wish to express my sincere thanks to the following electronics component manufacturers for providing material that was used in this book: Burr-Brown Corporation, Intersil Inc., Plessey Solid State, Supertex Inc., and Thomson Components—Mostek Corporation. These firms are identified in each figure according to the following key:

BB—Burr-Brown Corporation
IN—Intersil Inc.
PL—Plessey Solid State
SU—Supertex Inc.
TH—Thomson Components—Mostek Corporation

Preface

The science of electronics has made a quantum leap forward since the invention of the first simple integrated circuit only a generation ago. Today, complex electronic circuits capable of performing very sophisticated functions are paradoxically much simpler to construct than the older discrete-component circuits they have replaced. This is because an integrated circuit is both a single component and, at the same time, an entire electronic circuit or even a large group of circuits and stages, containing tens, hundreds, or even thousands of separate discrete components.

We have only begun to scratch the surface for the possible uses of all of the many thousands of existing integrated circuits. The months and years ahead promise the introduction of new ICs, which will have capabilities that will be far beyond those that we have today. It can be truly said that in electronics research and development we have already reached the twenty-first century.

Introduction

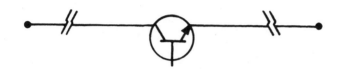

The purpose of this book is to provide information about practical electronic circuit devices and their applications. This book attempts to fill the gap between the electronics circuit books and the electronics project books. The former usually contain only electronics circuits and the latter almost always contain only information about very specific projects. While both of these types of books are very valuable, it was felt that there was a very definite need for a book that would provide *both* a wealth of practical electronics circuits *and* additional information about specific components. This is the type of practical information that electronics hobbyists, experimenters, technicians, and engineers will find very useful as a reference for electronics circuit design.

There are over 500 practical electronics circuits in this book. Many of the circuits are accompanied by descriptive text and other technical data. Interspersed among the circuits are more complete technical data on some of the popular types of devices, such as the operational amplifier, instrumentation amplifier, and the very popular 555 and 556 timers. This is the type of material that is needed to take an electronics circuit diagram and to convert it into a working electronic device or project.

The chapters of this book are organized according to different electronics applications. At the top of most of the pages, the major component numbers are shown. Many components can be used in several different applications. Op amps, for example (which are covered in Chapter 2), are also used for other applications in other chapters. For instance, an op amp can be used as a simple buffer amplifier (Chapter 2) and in an oscillator circuit (Chapter 3). Throughout the book you will find cross-references to other circuits in other chapters. The extensive index provides yet another aid to locating both electronics circuits and electronic devices. The Appendix gives a listing of suppliers of electronic parts and components. The Acknowledgments provides a key for original illustration suppliers.

1
Basic Electronic Circuits

SWITCH MODE POWER SUPPLY

The TEA2018A is a large diffusion low-cost integrated circuit packaged in an 8-pin mini-Dip CB-98 case and designed for the control of switch mode power supplies in fly-back discontinuous mode.

By addition of external switching transistor, power levels of as high as 90 W are efficiently handled.

Application areas include: video display units, into video games, T.V. sets, Hi-Fi amplifiers, function generators.

Where synchronization is required, use the TEA2019.

- ☐ Direct drive of the external switching transistor
- ☐ Positive and negative output currents of up to 0.5 A
- ☐ Current limitation
- ☐ Demagnetization sensing
- ☐ Full protection against overloads and short-circuits
- ☐ Output current is a function of the switching transistor collector current: $I_c = kI_B$ programmed externally
- ☐ Low standby current before starting
- ☐ $t_{on(min)}$: 2 μs
- ☐ Thermal protection

For further information refer to application note NA041.

General Description

The principles of the switching regulator described here is discontinuous mode with fixed frequency operation under normal regulating conditions.

However, lack of periods will occur if an overload or short-circuit is detected by demagnetization sensing circuitry implemented on the chip. In this case, any new cycle of operation is disabled until the secondary current has fallen to zero.

On every fall of the oscillator sawtooth, the flip-flop is set by a 2 μs pulse, thereby initiating the output current by supplying a large current pulse and thus providing a rapid turn on of the switching transistor. This current pulse is also used for $t_{on(min)}$ function.

Under normal operating conditions, the flip-flop is reset by a signal issued from comparing the following:

- ☐ The sawtooth waveform representing the collector current of the switching transistor, sampled across the emitter shunt resistor.
- ☐ The output of the error amplifier.

If the voltage drop across the shunt resistor is in excess of -1 V, the flip-flop is reset and as a result the output current is limited.

Outside the regulation area, and in the absence of current limitation, the flip-flop can be reset by a $t_{on(max)}$ signal of about 70% of the period.

TEA2018A
MAXIMUM RATINGS

Rating	Symbol	Value	Unit
Positive supply voltage	V_{CC}^+	15	V
Negative supply voltage	V_{CC}^-	−5	V
Output current	I_O	±0.5	A
Peak output current (Duty cycle < 5%)	I_O (peak)	±1	A
Pin 3 input current	$I_{(3)}$	±5	mA
Junction temperature	T_j	+150	°C
Operating ambient temperature range	T_{oper}	−20 to +70	°C
Storage temperature range	T_{stg}	−40 to +150	°C

THERMAL CHARACTERISTICS

Characteristic	Symbol	Value	Unit
Junction-ambient thermal resistance*	$R_{th(j-a)}$	80	°C/W

* Ex : 0.7 Watt in the I.C. makes its junction temperature grow 56°C over the ambient temperature.
To keep a good reliability a +100°C maximum junction working temperature is recommended.

BLOCK DIAGRAM

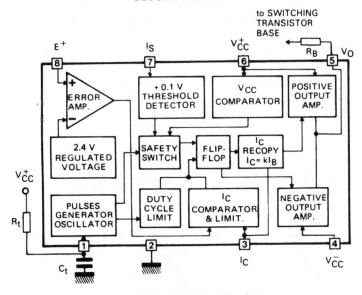

1 - C_t : Oscillator capacitor and resistor
2 - GND : Ground
3 - I_C : I_C sample (negative)
4 - V_{CC}^- : Negative supply (output stage)
5 - V_O : Output
6 - V_{CC}^+ : Positive supply voltage
7 - I_S : Demagnetization sensing
8 - E^+ : Error amplifier non-inverting input

Fig. 1-1

(TH)

TEA2018A

PIN ASSIGNMENT
(Top view)

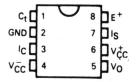

ELECTRICAL CHARACTERISTICS

$T_{amb} = +25°C$, potentials referenced to ground
(Unless otherwise specified)

Characteristic	Symbol	Min	Typ	Max	Unit
Positive supply voltage	V_{CC}^+	6	8	15	V
Negative auxiliary voltage	V_{CC}^-	−1	−3	−5	V
Release voltage of the supply voltage (V_{CC} increasing)	−	−	5.8	6.6	V
Stop voltage of the supply voltage (V_{CC} decreasing)	−	4.2	4.9	−	V
Hysteresis on V_{CC} threshold	$\Delta V_{(2-6)}$	0.6	0.9	1.5	V
Standby supply current before starting V_{CC}^+/GND < +5.8 V	$I_{(2-6)}$	−	1	1.6	mA
Current limitation threshold	I_C	−1080	−980	−880	mV
Current sample input resistance	$R_{(7)}$	−	1000	−	Ω
Demagnetization sensing threshold	I_S	75	100	125	mV
Demagnetization sensing input current ($V_{(2-7)} = 0$)	$I_{(7)}$	−	1	−.	μA
Maximum conducting duty cycle	−	60	70	−	%T_o
Error amplifier gain	−	−	50	−	−
Error amplifier input current	$I_{(8)}$	−	2	−	μA
Internal reference voltage	V_{ref}	2.3	2.4	2.5	V
Reference voltage temperature drift	−	−	10^{-4}	−	V/°C
Oscillator frequency drift with temperature ($V_{CC} = +8$ V)	−	−	0.005	−	%/°C
Oscillator frequency drift with V_{CC} (6 V < V_{CC} < 15 V)	−	−	0.2	−	%/V

Fig. 1-2

(TH)

In order to save power, the positive base current after the starting pulse becomes an increasing function of the collector current (this current is sampled across the emitter shunt resistor). The $\dfrac{I_C}{I_B}$ can be programmed as follows:

$$\frac{I_C}{I_B} = \frac{R_B}{R_e}$$

R_e must be calculated so as to obtain a 1 V voltage drop across it for the value of the limiting current.

Then R_B is chosen to give the required forced gain.

TEA2018A

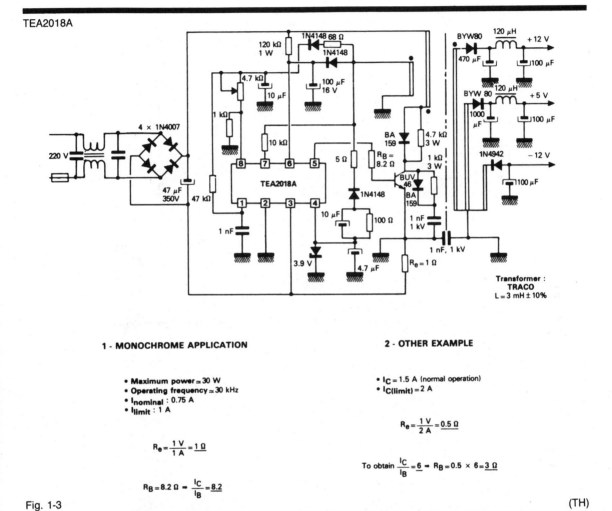

1 - MONOCHROME APPLICATION

- Maximum power ≃ 30 W
- Operating frequency ≃ 30 kHz
- $I_{nominal}$: 0.75 A
- I_{limit} : 1 A

$$R_e = \frac{1\,V}{1\,A} = \underline{1\,\Omega}$$

$$R_B = 8.2\,\Omega \Rightarrow \frac{I_C}{I_B} = \underline{8.2}$$

2 - OTHER EXAMPLE

- $I_C = 1.5$ A (normal operation)
- $I_{C(limit)} = 2$ A

$$R_e = \frac{1\,V}{2\,A} = \underline{0.5\,\Omega}$$

To obtain $\frac{I_C}{I_B} = 6 \Rightarrow R_B = 0.5 \times 6 = \underline{3\,\Omega}$

Fig. 1-3

(TH)

When the positive base current is removed, 1 μs will elapse before the application of negative base current therefore allowing a rapid fall of the collector current.

Pin 4 (V_{CC}) must be supplied by a negative voltage of −2 V to −3 V.

Starting Process. Prior to starting, a low current is drawn from +300 V through a high value resistor.

This current will charge the power supply storage capacitor of the IC. No output pulse will be available before the voltage across the capacitor has reached 6 V. During this time, the device will draw only 1 mA. When the voltage across the capacitor reaches 6 V, base current pulses will appear at the output. The energy drawn by these pulses will tend to discharge the power supply storage capacitor. However a hysteresis of about 1 V is implemented to allow a satisfactory operation even with 5 V. Then the auxiliary winding of the transformer will provide the power required by the IC.

TEA2018A

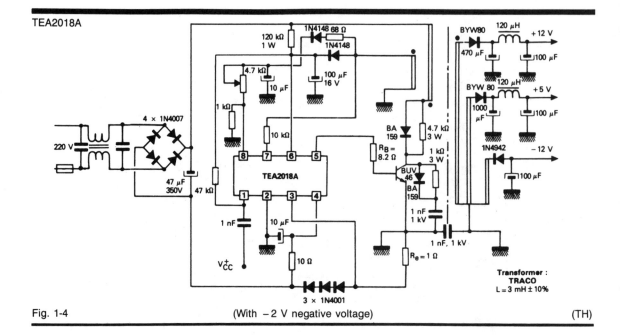

Fig. 1-4 (With − 2 V negative voltage) (TH)

ADJUSTABLE HIGH VOLTAGE SUPPLY

This circuit was developed to provide the high voltage for a small CRT vector graphics display. Previous similar designs using bipolar transistors require more base current than can be provided by an inexpensive operational amplifier.

The high voltage supply employs a VN1116N2 FET connected to the windings of the Murata TV flyback transformer to implement a blocking oscillator. Other flyback transformers may be used if they have a suitable separate winding to provide a pulse of about 10 volts to drive the gate of the FET. The .01 μF and the .0018 μF capacitors are needed to suppress high frequency oscillations and to give a better pulse shape to improve rectifier efficiency.

A useful unique feature of the oscillator arises from the fact that the degree of core saturation and the resulting high voltage are easily adjusted over a wide range by varying the bias voltage applied to the FET gate. The bias voltage can be provided easily by an inexpensive operational amplifier, resulting in a very simple adjustable and regulated high voltage. This can be accomplished using one amplifier input derived from the focus voltage output of the flyback transformer and the other input voltage varied using a potentiometer. Output voltages from 3 to 12 kV can be obtained.

VN1116N2

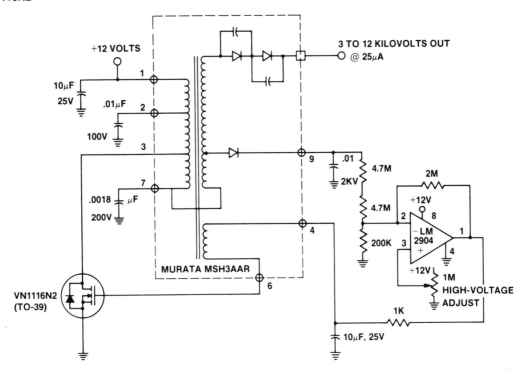

Fig. 1-5

(SU)

HIGH VOLTAGE VARIABLE SUPPLY

Pass transistors in high voltage high current service, require large insulated heatsinks that endanger safe and reliable operation. VP03's large SOA and grounded drain terminal make this a practical circuit, with the simplest heatsink requirements (no other component requires heatsinking). Several VP03 pass transistors will directly parallel-connect in this circuit to enhance output rating.

VP0104N3, VN0545N3, VP0345N1

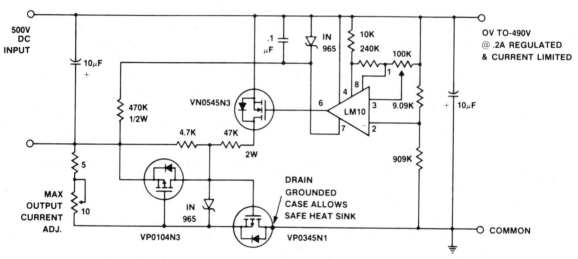

Fig. 1-6

(SU)

250 KHZ SWITCH-MODE 5 VDC SUPPLY

The VP1204N5 is coupled directly to the control IC to illustrate a compact design. The switching regulator takes advantage of the body diode in the VP12, which acts as a free wheeling diode for limiting inductive voltages. This eliminates the need for an external and costly schottky diode.

VP1204N5

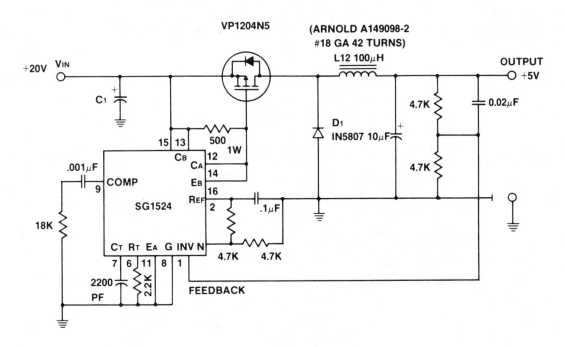

Fig. 1-7

(SU)

200 WATT OFF-LINE FLYBACK SUPPLY

The output transistor Q1 receives the oscillator drive until an output-overvoltage turns on the isolator. This actuates the Q2 clamp transistor, which turns off the oscillator. The 0.22 ohm source resistor generates automatic clamping oscillator signals at around 5 amperes transformer current. Q1's gate is powered simply by the 311 IC.

Switcher operating frequency begins with the time constants contained within the oscillator circuit, but as output error voltage minimizes producing shorter error signals, oscillator frequency rises and is dependent more on the system time constants. Replacing the oscillator with a 555 type astable with reset controlled circuit will stabilize the switcher frequency.

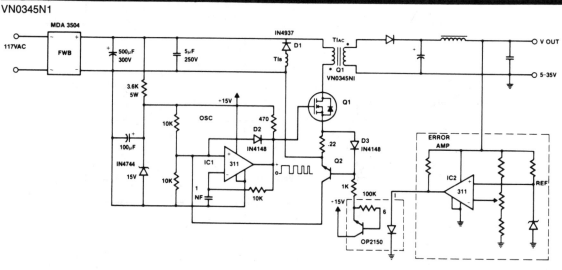

Fig. 1-8

(SU)

LM111

LOW VOLTAGE ADJUSTABLE REFERENCE SUPPLY

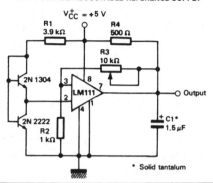

Fig. 1-9 (TH)

LM105, LM205, LM305

LINEAR REGULATOR WITH FOLDBACK CURRENT LIMITING   **SWITCHING REGULATOR**

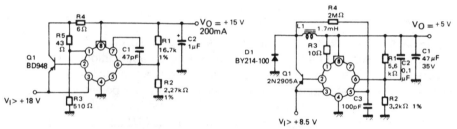

10 A REGULATOR WITH FOLDBACK CURRENT LIMITING   **1 A REGULATOR WITH PROTECTIVE DIODES**

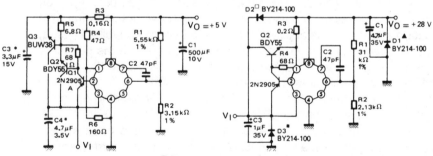

□ Protects against shorted input or inductive loads on unregulated supply.

■ Protects against input voltage reversal.

▲ Protects against output voltage reversal.

* Solid tantalum

CURRENT REGULATOR   **SHUNT REGULATOR (V$_O$<0)**

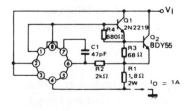
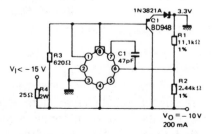

Fig. 1-10 (TH)

ADJUSTABLE REGULATOR

In operation, the LM338 develops a nominal 1.25 V reference voltage, $V_{(ref)}$, between the output and adjustment terminal. The reference voltage is impressed across program resistor R1 and, since the voltage is constant, a constant current I_1 then flows through the output set resistor R2, giving an output voltage of

$$V_O = V_{(ref)}(1 + \frac{R2}{R1}) + I_{adj}R2$$

Since the 50 μA current from the adjustment terminal represents an error term, the LM338 was designed to minimize I_{adj} and make it very constant with line and load changes. To do this, all quiescent operating current is returned to the output establishing a minimum load current requirement. If there is insufficient load on the output, the output will rise.

External Capacitors

An input bypass capacitor is recommended. A 0.1 μF disc or 1 μF solid tantalum on the input is suitable input bypassing for almost all applications. The device is more sensitive to the absence of input bypassing when adjustment or output capacitors are used but the above values will eliminate the possibility of problems.

The adjustment terminal can be bypassed to ground on the LM338 to improve ripple rejection. This bypass capacitor prevents ripple from being amplified as the output voltage is increased. With a 10 μF bypass capacitor 75 dB ripple rejection is obtainable at any output level. Increases over 20 μF do not appreciably improve the ripple rejection at frequencies above 120 Hz. If the bypass capacitor is used, it is sometimes necessary to include protection diodes to prevent the capacitor from discharging through internal low current paths and damaging the device.

In general, the best type of capacitors to use are solid tantalum. Solid tantalum capacitors have low impedance even at high frequencies. Depending upon capacitor construction, it takes about 25 μF in aluminum electrolytic to equal 1 μF solid tantalum at high frequencies. Ceramic capacitors are also good at high frequencies, but some types have a large decrease in capacitance at frequencies around 0.5 MHz. For this reason, 0.01 μF disc may seem to work better than a 0.1 μF disc as a bypass.

Although the LM338 is stable with no output capacitors, like any feedback circuit, certain values of external capacitance can cause excessive ringing. This occurs with values between 500 pF and 5000 pF. A 1 μF solid tantalum (or 25 μF aluminum electrolytic) on the output swamps this effect and ensures stability.

Load Regulation

The LM338 is capable of providing extremely good load regulation but a few precautions are needed to obtain maximum performance. The current set resistor connected between the adjustment terminal and the output terminal (usually 240 Ω) should be tied directly to the output of the regulator rather than near the load.

LM138, LM238

+1.2 V to +25 V ADJUSTABLE REGULATOR

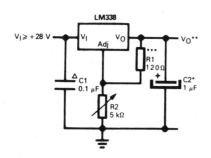

Fig. 1-11 (TH)

LM138, LM238

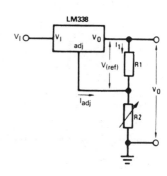

Fig. 1-12 (TH)

LM138, LM238, LM338

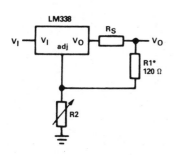

Regulator with line resistance in output lead

Fig. 1-13 (TH)

LM138, LM238, LM338

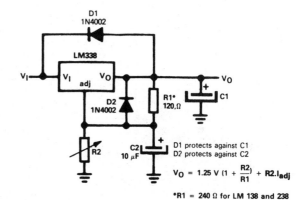

D1 protects against C1
D2 protects against C2

$$V_O = 1.25\ V\ (1 + \frac{R2}{R1}) + R2.I_{adj}$$

*R1 = 240 Ω for LM 138 and 238

Fig. 1-14 Regulator with protection diodes (TH)

This eliminates line drops from appearing effectively in series with the reference and degrading regulation. For example, a 15 V regulator with 0.05 Ω resistance between the regulator and load will have a load regulation due to line resistance of 0.05 Ω × I_L. If the set resistor is connected near the load the effective line resistance will be 0.05 Ω (1 + R2/R1) or in this case, 11.5 times worse.

Figure 1-13 shows the effect of resistance between the regulator and 140 Ω set resistor.

With the TO-3 package, it is easy to minimize the resistance from the case to the set resistor, by using 2 separate leads to the case. The ground of R2 can be returned near the ground of the load to provide remote ground sensing and improve load regulation.

LM138, LM238, LM338

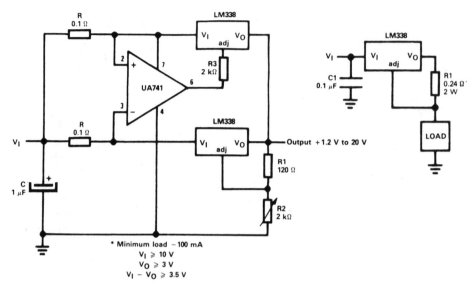

10A REGULATOR

5A CURRENT REGULATOR

* Minimum load – 100 mA
$V_I \geqslant 10\ V$
$V_O \geqslant 3\ V$
$V_I - V_O \geqslant 3.5\ V$

15 A REGULATOR

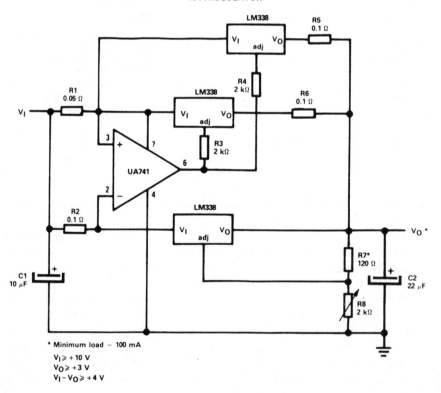

* Minimum load – 100 mA
$V_I \geqslant +10\ V$
$V_O \geqslant +3\ V$
$V_I - V_O \geqslant +4\ V$

Fig. 1-15

(TH)

LM138, LM238, LM338

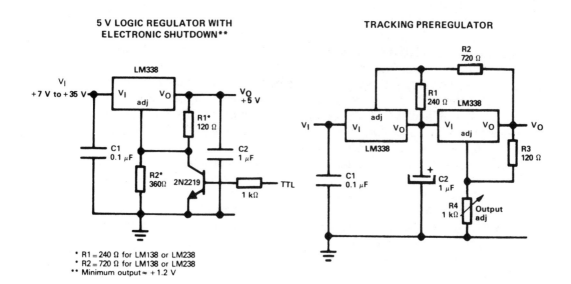

5 V LOGIC REGULATOR WITH ELECTRONIC SHUTDOWN**

TRACKING PREREGULATOR

* R1 = 240 Ω for LM138 or LM238
* R2 = 720 Ω for LM138 or LM238
** Minimum output ≈ +1.2 V

SLOW TURN-ON 15 V REGULATOR

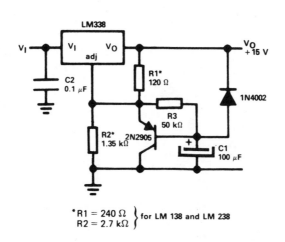

*R1 = 240 Ω } for LM 138 and LM 238
R2 = 2.7 kΩ }

Fig. 1-16   (TH)

Protection Diodes

When external capacitors are used with any IC regulator it is sometimes necessary to add protection diodes to prevent the capacitors from discharging through low current points into the regulator. Most 20 μF capacitors have low enough internal series resistance to deliver 20 A spikes when shorted. Although the surge is short, there is enough energy to damage parts of the IC.

When an output capacitor is connected to a regulator and the input is shorted, the output capacitor will discharge into the output of the regulator. The discharge current depends on the value of the capacitor, the output voltage of the regulator, and the rate of decrease of V_I. In the LM338 this discharge path is through a large junction that is able to sustain 25 A surge with no problem. This is not true of other types of positive regulators. For output capacitors of 100 μF or less at output of 15 V or less, there is no need to use diodes.

The bypass capacitor on the adjustment terminal can discharge through a low current junction. Discharge occurs when either the input or output is shorted. Internal to the LM338 is a 50 Ω resistor which limits the peak discharge current. No protection is needed for output voltages of 25 V or less and 10 μF capacitance. Figure 1-14 shows an LM338 with protection diodes included for use with outputs greater than 25 V and high values of output capacitance.

CMOS PROGRAMMABLE MICROPOWER VOLTAGE REGULATORS

The ICL7663 (positive) and ICL7664 (negative) series regulators are low-power, high-efficiency devices which accept inputs from 1.6 V to 10 V and provide adjustable outputs over the same range at currents up to 40 mA. Operating current is typically less than 4 μA, regardless of load.

Output current sensing and remote shutdown are available on both devices, thereby providing protection for the regulators and the circuits they power. A unique feature, on the ICL7663 only, is a negative temperature coefficient output. This can be used, for example, to efficiently tailor the voltage applied to a multiplexed LCD through the driver (e.g., ICM7231/2/3/4) so as to extend the display operating temperature range many times.

The ICL7663 and ICL7664 are available in either an 8-pin plastic, TO-99 can, CERDIP, and SOIC packages.

Features

- ☐ Ideal for battery-operated systems: less than 4 μA typical current drain
- ☐ Will handle input voltages from 1.6 V to 16 V
- ☐ Very low input-output differential voltage
- ☐ 1.3 V bandgap voltage reference
- ☐ Up to 40 mA output current
- ☐ Output shutdown via current-limit sensing or external logic signal
- ☐ Output voltages programmable from 1.3 V to 16 V
- ☐ Output voltages with programmable negative temperature coefficients (ICL7663 only)

ICL7663/7664

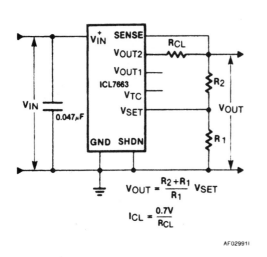

$$V_{OUT} = \frac{R_2 + R_1}{R_1} V_{SET}$$

$$I_{CL} = \frac{0.7V}{R_{CL}}$$

AF02991I

Basic application of ICL7663 as positive regulator with current limit

$$V_{OUT} = \frac{R_2 + R_1}{R_1} V_{SET}$$

$$I_{CL} = \frac{0.35V}{R_{CL}}$$

AF02890I

Basic application of ICL 7664 as negative regulator with current limit

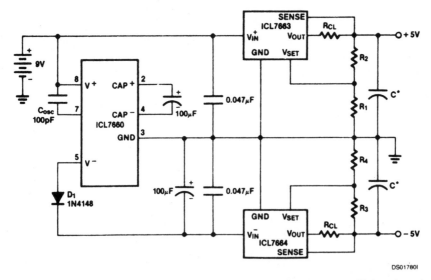

DS01780I

*Values depend on load characteristics

Generating regulated split supplies from a single supply

The oscillation frequency of the ICL7660 is reduced by the external oscillator capacitor, so that it inverts the battery voltage more efficiently.

Fig. 1-17

ICL7663/7664

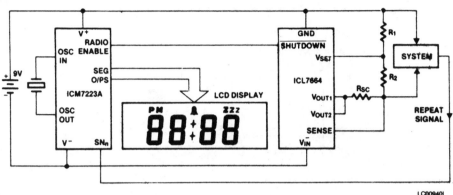

Once a day system

This circuit will turn on a regulated supply to a system for one minute every day, via the $\overline{\text{SHUTDOWN}}$ pin on the ICL7664, and under control of the ICM7223A Alarm Clock circuit. If the system decides it needs another one minute activation, pulling the REPEAT line to V$^+$ (GND) during one activation will trigger a subsequent activation after a snooze interval set by the choice of SN pins (2 mins shown). Alternatively, activation of the Sleep timer, without pause, can be achieved. See ICM7223A data sheet for details.

Fig. 1-18   (IN)

LM137, LM237, LM337

TYPICAL APPLICATIONS

ADJUSTABLE NEGATIVE VOLTAGE REGULATOR

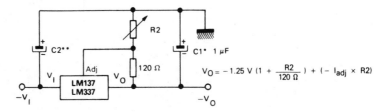

$$V_O = -1.25 \text{ V} (1 + \frac{R2}{120 \text{ }\Omega}) + (-I_{adj} \times R2)$$

* C1 = 1 μF solid tantalum or 10 μF aluminium electrolytic required for stability
** C2 = 1 μF solid tantalum is required only if regulator is more than 10 cm from power supply filter capacitor

ADJUSTABLE LAB VOLTAGE REGULATOR

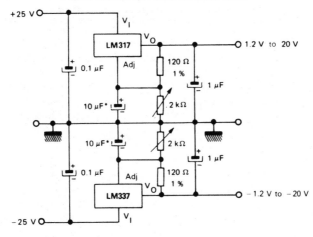

* The 10 μF capacitors are optional to improve ripple rejection

CURRENT REGULATOR

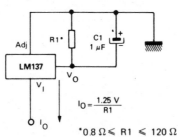

$$I_O = \frac{1.25 \text{ V}}{R1}$$

*0.8 Ω ≤ R1 ≤ 120 Ω

Fig. 1-19

(TH)

LM137, LM237, LM337

NEGATIVE REGULATOR WITH PROTECTION DIODES

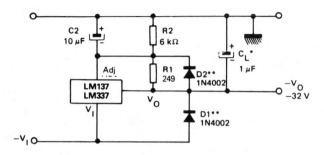

* When C_L is larger than 20 μF, D1 protects the LM 137 in case the input supply is shorted

**When C2 is larger than 10 μF and V_O is larger than -25 V, D2 protects the LM 137 in case the output is shorted

* -5.2 V REGULATOR WITH ELECTRONIC SHUTDOWN

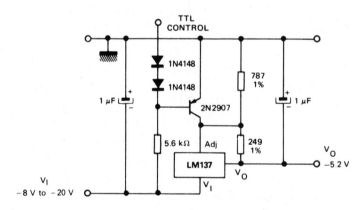

* Minimum output $\cong -1.3$ V when control input is low

ADJUSTABLE CURRENT REGULATOR

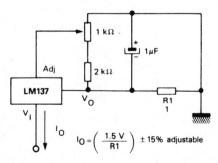

$$I_O = \left(\frac{1.5 \text{ V}}{R1} \right) \pm 15\% \text{ adjustable}$$

Fig. 1-20

(TH)

LM117, LM217, LM317   "Laboratory" power supply with adjustable current limit and output voltage

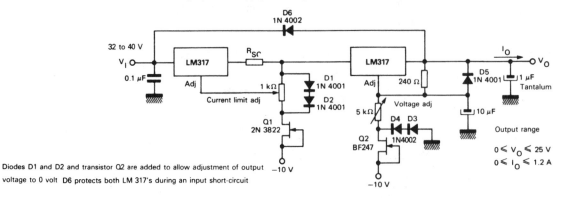

Diodes D1 and D2 and transistor Q2 are added to allow adjustment of output
voltage to 0 volt D6 protects both LM 317's during an input short-circuit

Output range

$0 \leqslant V_O \leqslant 25$ V
$0 \leqslant I_O \leqslant 1.2$ A

Electronic shut down regulator   Slow turn-on regulator   Adjustable current limiter

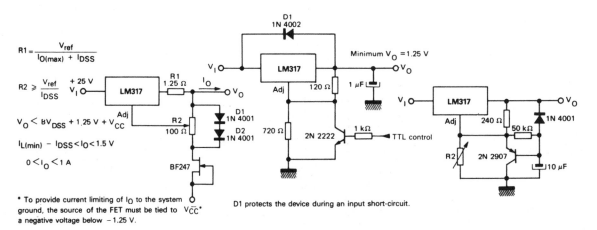

$$R1 = \frac{V_{ref}}{I_{O(max)} + I_{DSS}}$$

$$R2 \geqslant \frac{V_{ref}}{I_{DSS}}$$

$$V_O < BV_{DSS} + 1.25\ V + V_{CC}$$

$$I_{L(min)} - I_{DSS} < I_O < 1.5\ V$$

$$0 < I_O < 1\ A$$

* To provide current limiting of I_O to the system
ground, the source of the FET must be tied to $V_{\overline{CC}}$*
a negative voltage below -1.25 V.

D1 protects the device during an input short-circuit.

Fig. 1-21

1.2 to 25 V adjustable regulator   Precision current limiter

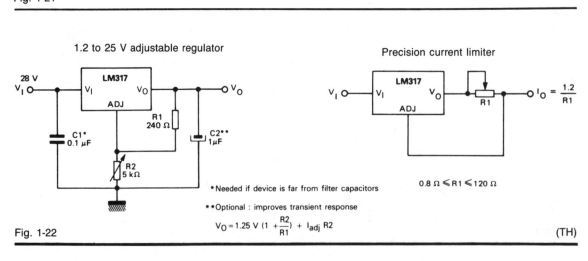

*Needed if device is far from filter capacitors

**Optional : improves transient response

$$V_O = 1.25\ V\ (1 + \frac{R2}{R1}) + I_{adj}\ R2$$

$$0.8\ \Omega \leqslant R1 \leqslant 120\ \Omega$$

Fig. 1-22   (TH)

LM109, LM209, LM309

☆ HIGH STABILITY REGULATOR

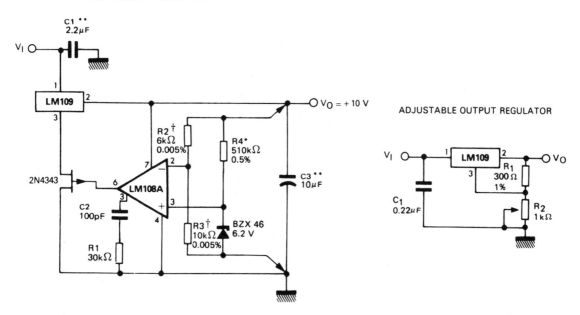

ADJUSTABLE OUTPUT REGULATOR

☆ Regulation better than 0.01% load, line and temperature can be obtained

* Determines zener current. May be adjusted to minimize thermal drift

** Solid tantalum

† High stability resistors

FIXED 5 V REGULATOR

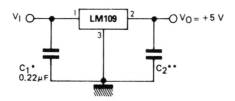

* Required if regulator is located an appreciable distance from power supply filter capacitor.

** Although no output capacitor is needed for stability, it does improve transient response.

CURRENT REGULATOR

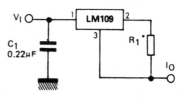

* Determines output current

Fig. 1-23

L200

POSITIVE AND NEGATIVE REGULATOR

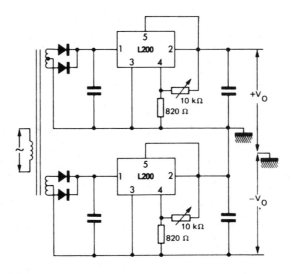

TRACKING VOLTAGE REGULATOR

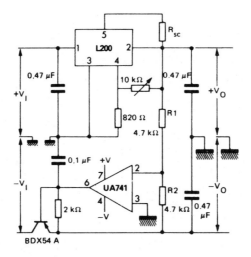

TEST CIRCUIT FOR PULSE MEASUREMENTS

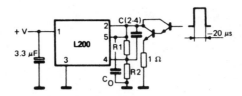

Fig. 1-24 (TH)

L200

PROGRAMMABLE CURRENT REGULATOR

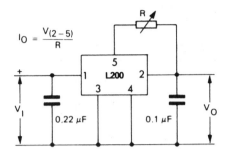

$$I_O = \frac{V_{(2-5)}}{R}$$

PROGRAMMABLE VOLTAGE REGULATOR

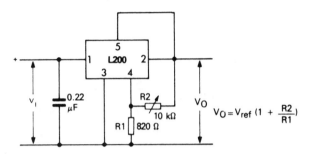

$$V_O = V_{ref} \left(1 + \frac{R2}{R1}\right)$$

PROGRAMMABLE VOLTAGE REGULATOR
WITH CURRENT LIMITING

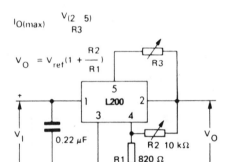

$$I_{O(max)} \quad \frac{V_{(2\;5)}}{R3}$$

$$V_O = V_{ref}\left(1 + \frac{R2}{R1}\right)$$

SWITCHING REGULATOR

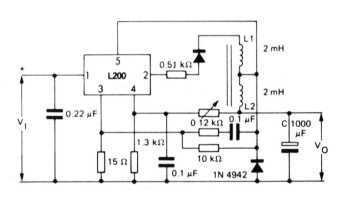

HIGH CURRENT VOLTAGE REGULATOR WITH
SHORT CIRCUIT PROTECTION

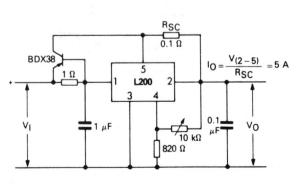

$$I_O = \frac{V_{(2-5)}}{R_{SC}} = 5\;A$$

DIGITALLY SELECTED REGULATOR
WITH INHIBIT

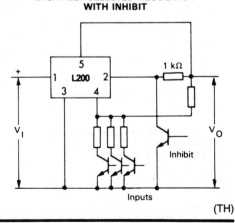

Fig. 1-25

(TH)

LM123, LM223, LM323

BASIC 3 A REGULATOR

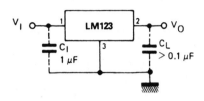

TRIMMING OUTPUT TO 5 V

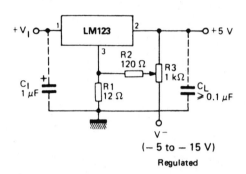

V^-
$(-5 \text{ to } -15 \text{ V})$
Regulated

C_I = Required if regulator is distant
   from filter capacitors

C_L = Regulator is stable with no load
   capacitor into resistive loads.

10 A REGULATOR WITH COMPLETE OVERLOAD PROTECTION

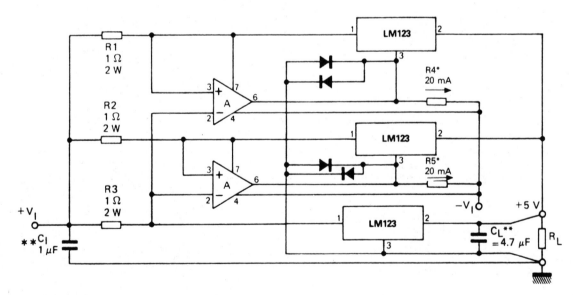

* Selected for 20 mA current from unregulated negative supply
** Solid tantalum
A = LM101A, LM201A, LM301A

Fig. 1-26

(TH)

LM123, LM223, LM323

ADJUSTABLE REGULATOR 0 – 10 V / 3 A

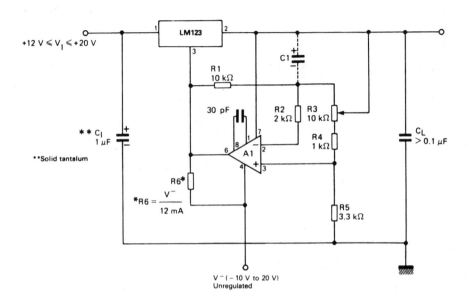

A1 = LM 101 A, LM 201 A, LM 301 A.
C1 = 2 μF optional - improves ripple rejection, noise and transient response.

Fig. 1-27   (TH)

UA723

BASIC CIRCUITS

FOLDBACK CURRENT LIMITING

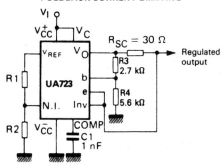

Regulated output voltage	+ 5 V
Line regulation (ΔV_I = 3 V)	0.5 mV
Load regulation (ΔI_L = 10 mA)	1 mV
Short-circuit current	20 mV

POSITIVE FLOATING REGULATOR
V_I = + 85 V

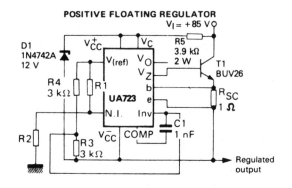

Regulated output voltage	+ 50 V
Line regulation (ΔV_I = 20 V)	15 mV
Load regulation (ΔI_L = 50 mA)	20 mV

BASIC LOW VOLTAGE REGULATOR
(V_O = 2 to 7 V)

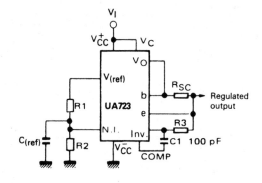

Regulated output voltage	5 V
Line regulation (ΔV_I = 3 V)	0.5 mV
Load regulation (ΔI_L = 50 mA)	1.5 mV

NOTE 3 : $R3 = \dfrac{R1\ R2}{R1\ +\ R2}$ for minimum temperature drift

BASIC HIGH VOLTAGE REGULATOR
(V_O = 7 to 37 V)

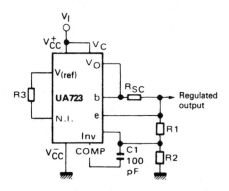

Regulated output voltage	15 V
Line regulation (ΔV_I = 3 V)	1.5 mV
Load regulation (ΔI_L = 50 mA)	4.5 mV

NOTE : $R3 = \dfrac{R1\ R2}{R1\ +\ R2}$ for minimum temperature drift

R3 may be eliminated for minimum component count

Fig. 1-28

UA723

NEGATIVE VOLTAGE REGULATOR (Note 1)

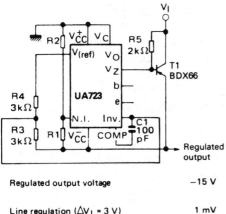

Regulated output voltage	−15 V
Line regulation (ΔV_I = 3 V)	1 mV
Load regulation (ΔI_L = 100 mA)	2 mV

Note 1 : For applications using TO-100 metal cans ; V_Z can be implemented externally by connecting a 6.2 V zener diode to V_O pin.

POSITIVE VOLTAGE REGULATOR
(External NPN Pass Transistor)

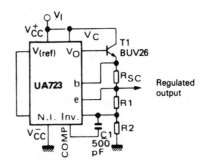

Regulated output voltage	+ 15 V
Line regulation (ΔV_I = 3 V)	1.5 mV
Load regulation (ΔI_L = 1 A)	15 mV

POSITIVE VOLTAGE REGULATOR
(External PNP Pass Transistor)

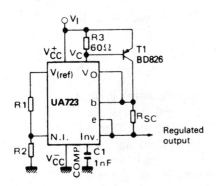

Regulated output voltage	+ 5 V
Line regulation (ΔV_I = 3 V)	0.5 mV
Load regulation (ΔI_L = 1 A)	5 mV

SHUNT REGULATOR

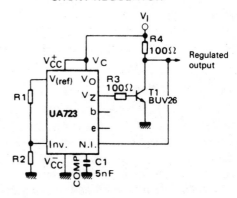

Regulated output voltage	+ 5 V
Line regulation (ΔV_I = 10 V)	0.5 mV
Load regulation (ΔI_L = 100 mA)	1.5 mV

Fig. 1-29

(TH)

BATTERY POWERED REGULATOR FOR ANALOG ICs

This simple, high-performance power supply for battery operated equipment is as efficient as a good "switcher" design, but doesn't have the radiated field and ripple normally associated with switchers. The output voltage (6 volts) is high enough to power a wide variety of integrated circuits.

A 2.5 volt reference is furnished for external use. Since most op amps work closer to the negative rail than to the positive rail, 2.5 volts is a better biasing voltage than 3.0 volts that would be obtained by "splitting the rails." The reference can sink 10 mA and can source 50 μA. If more source current is needed, it can be provided by decreasing the value of the reference biasing resistor (R1).

Advantages of Using a MOSFET

The regulator shown here was originally designed using a pnp pass transistor. In that form, the design offered enough advantages over integrated circuit regulators and it was put into production. Later, the possibility of improving the design by using an Enhancement Mode p-channel MOSFET as the pass element was considered. After finding a source of p-channel units with sufficiently low threshold voltage (SUPERTEX, INC.) the design was revised to accommodate a MOSFET, and tested. The results were as follows:

☐ somewhat lower quiescent current.
☐ voltage burden (input-output differential) substantially reduced.
☐ output impedance much lower.
☐ virtually free of negative input resistance effects that plague low voltage burden regulators which use a bipolar transistor as the pass element.
☐ the gate capacitance of the MOSFET was put to use in an improved phase compensation scheme that virtually eliminated load reactance constraints and made regulation much faster.
☐ no performance tradeoffs were required. The MOSFET design was equal to or superior to the bipolar design in every respect.

Circuit Description

Biased by R1, active zener Q1 provides a 2.5 volt reference at the emitter of sense transistor Q2. Voltage divider D1, D2, R2/R3 divides the 6 volt rail down to 2 volts to furnish a sense voltage to the base of Q2. If the rail is less than 6.0 volts, the sense voltage drops below 2.0 volts, increasing the forward bias of the base-emitter junction of Q2. The resultant increase in collector current is amplified by Q3, resulting in an increase of the gate source bias voltage of Q4. The drain current of Q4 increases, restoring the 6-volt rail to its proper voltage.

Diodes D1, D2 provide temperature compensation for the base-emitter junction of Q2. Capacitor C1 eliminates Miller rolloff of Q2 and speeds up the feedback path through the voltage divider. The gate capacitance of Q4 minimizes Miller rolloff of Q3, and establishes in conjunction with R5 a frequency rolloff at about 10 kHz. This rolloff makes it possible to provide dominant-pole phase compensation using

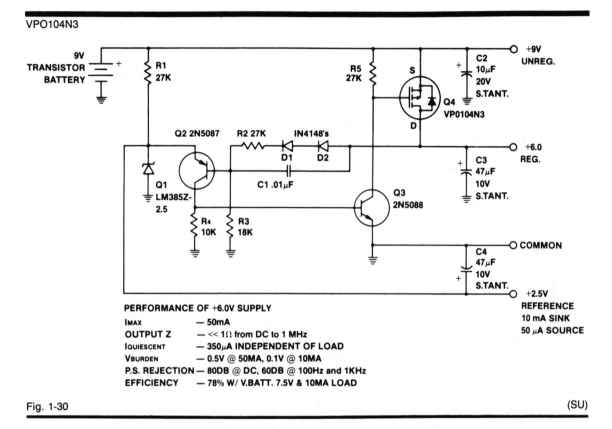

VPO104N3

PERFORMANCE OF +6.0V SUPPLY

I_{MAX}	— 50mA
OUTPUT Z	— << 1Ω from DC to 1 MHz
I_{QUIESCENT}	— 350µA INDEPENDENT OF LOAD
V_{BURDEN}	— 0.5V @ 50MA, 0.1V @ 10MA
P.S. REJECTION	— 80DB @ DC, 60DB @ 100Hz and 1KHz
EFFICIENCY	— 78% W/ V.BATT. 7.5V & 10MA LOAD

Fig. 1-30   (SU)

an inexpensive capacitor C2 at the output. Capacitor C3 inhibits parasitic oscillations and improves transient response, and C4 keeps the impedance of the voltage reference low even at high frequencies.

CURRENT LIMITED ± 5 V REGULATOR

DMOS SOA and transfer characteristics allow this circuit to sustain short circuit current safely until thermal switches disconnect input power. Maximum current is limited very simply by taking advantage of DMOS's VGS proportional to ID. The VP1204 (Q1) gate drive is limited to 8 volts, by the voltage divider, passing only about 10 amps. The input voltage drops 2-3 volts @ 10 A. Q2 is likewise controlled by a 4.5 V limited gate. This circuit exhibits fast load regulation, reducing the need for large output filter capacitors.

VN1204N1, VP1204N2

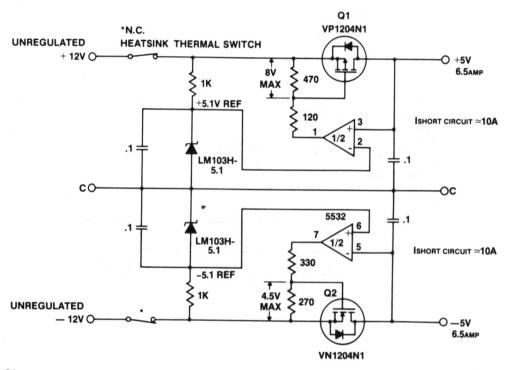

Fig. 1-31

(SU)

CURRENT PROTECTED PRECISION REGULATOR

Usually bipolar transistors are seen driving MOS power devices (advantageous with switches). The reverse scheme in this linear regulator, enables maximum output current control with two additional low level components: R sense & LM103-5.1. The pnp pass transistor is specified with min/max H_{FE}, making known the normal range of base current needed to support the designed load. As more load current also increases pnp base requirements and R sense voltage, there comes a point where the VN01's voltage-clamped gate simply limits further increases in pnp base current, limiting output current.

VN0104N5

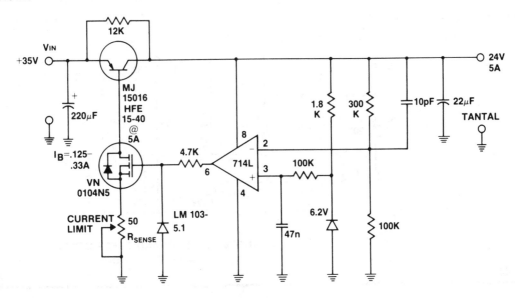

Fig. 1-32

(SU)

SHUNT VOLTAGE REGULATOR

This simple circuit, to handle a wide range of voltages, allows compact construction due to low component count. Voltage and current requirements will determine choice of Q1 and R4. R_D and zener D1 are necessary for safe operation of the 8211M IC when input voltage exceeds 30 V.

MOSFET

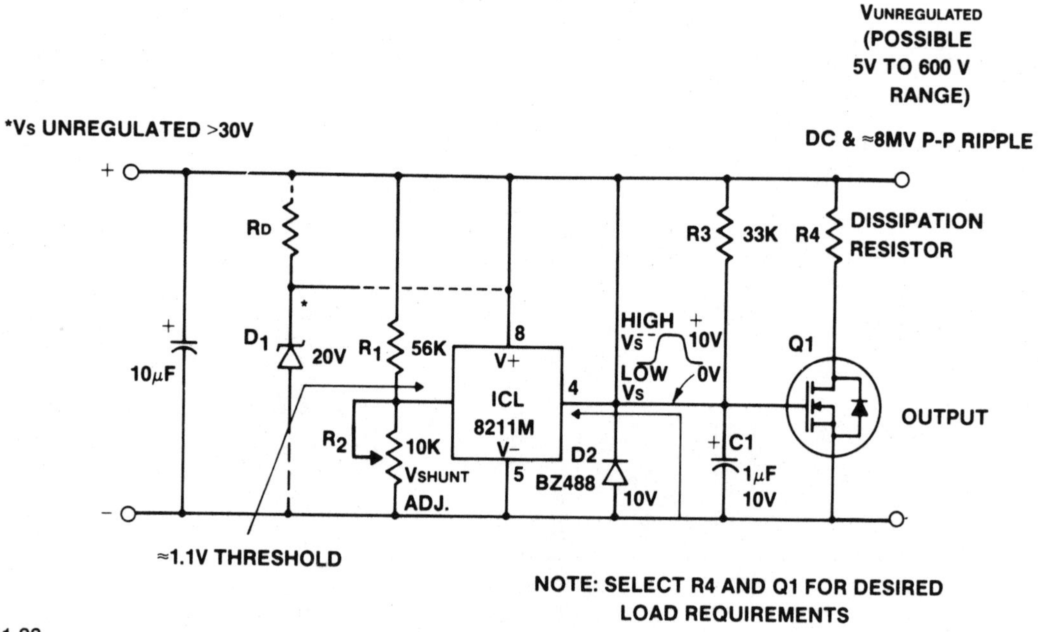

VUNREGULATED
(POSSIBLE
5V TO 600 V
RANGE)

*Vs UNREGULATED >30V

DC & ≈8MV P-P RIPPLE

≈1.1V THRESHOLD

NOTE: SELECT R4 AND Q1 FOR DESIRED
LOAD REQUIREMENTS

Fig. 1-33

(SU)

LINE POWERED 20 A NI-CAD CHARGER

Small size and minimal heat generation make this off line, push pull battery charger attractive for compact systems.

The 1525 switching regulator control IC provides the oscillator, voltage sense and output predriver functions. Line isolation and a simplified gate drive are achieved by T2.

VN0345N1

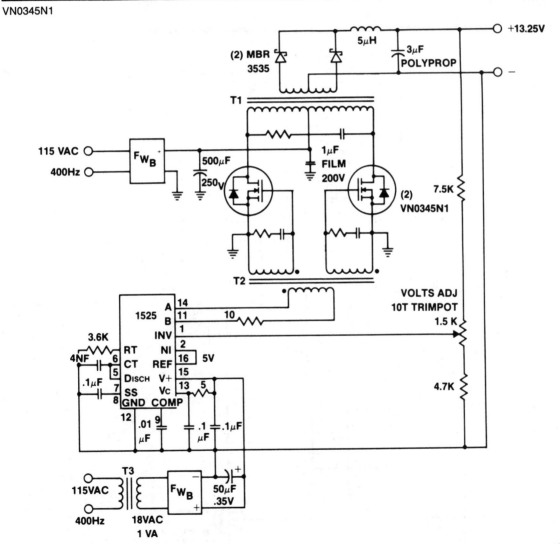

Fig. 1-34

(SU)

SWITCHING LEAD ACID BATTERY CHARGER

This is a 14.4 volts 1 amp current-limited battery charger. The circuit employs a step-down buck pulse width modulated switching regulator. Transistor Q2 and associated components act as a constant current source, providing a constant VGS to Q1, within the operating range. The VN0106N6 quad array (Q3 to Q6) performs several functions: System clock synch., power up control and power-in/logic level shifter. Q3 can synchronize the unit to a system clock to reduce switcher noise.

VN0106N6, VP1206N5
TL494

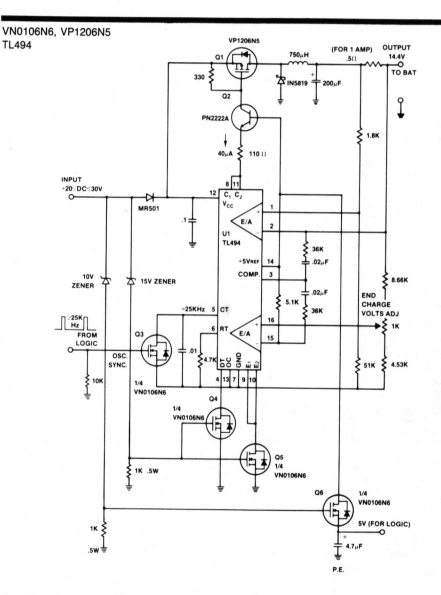

Fig. 1-35

OPA404

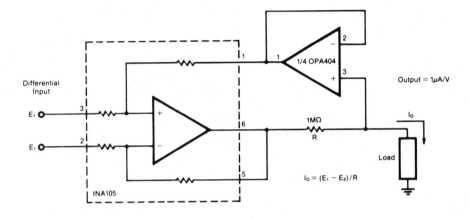

Output = 1μA/V

$I_o = (E_1 - E_2)/R$

Fig. 1-36   Voltage-controlled microamp current source.   (BB)

OPA404

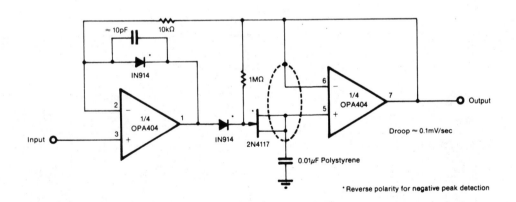

Droop ≈ 0.1mV/sec

0.01μF Polystyrene

* Reverse polarity for negative peak detection

Fig. 1-37   Low-droop positive peak detector.   (BB)

LM118, LM218, LM318

FAST SAMPLE AND HOLD

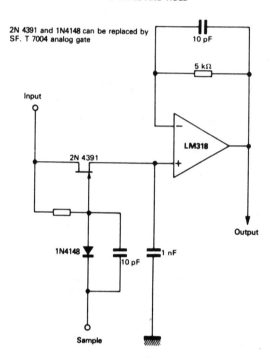

2N 4391 and 1N4148 can be replaced by
SF. T 7004 analog gate

Fig. 1-38

(TH)

LM101A, LM201A, LM301A

LM101A, LM201A, LM301A

FAST HALF WAVE RECTIFIER

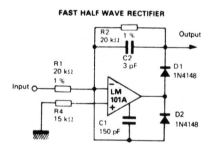

FAST AC/DC CONVERTER

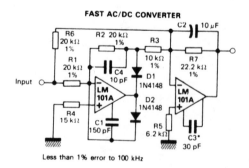

Less than 1% error to 100 kHz

Fig. 1-39   (TH) Fig. 1-40   (TH)

OPA156A

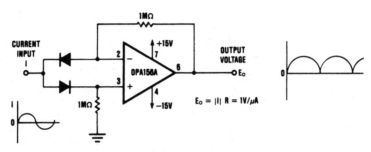

Absolute value current-to-voltage converter.

$$E_o = |I| R = 1V/\mu A$$

Fig. 1-41 (BB)

VFC32

Voltage-to-frequency converter at 500 kHz.

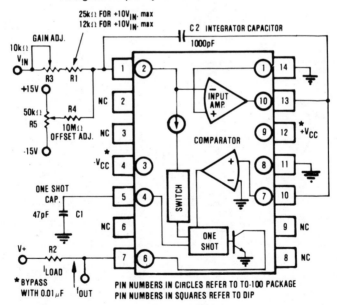

Connection diagram for V/F conversion, positive input voltages, 500 kHz maximum frequency.
For additional VFC32 circuits see Chapter 2 (Interfacing Circuits).

Fig. 1-42 (BB)

LM124

LOW DRIFT PEAK DETECTOR

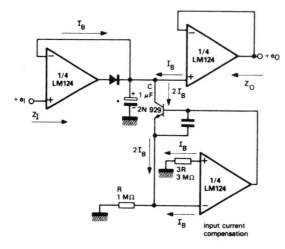

For more LM124 circuits see Chapter 2.

Fig. 1-43 (TH)

OPA600, 3553, 74LS27, AM686

NOTE: Power supply connections not shown.
All devices should be bypassed with 1—10μF capacitors at supply pins.

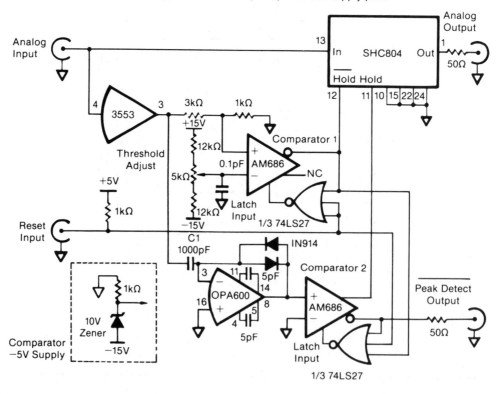

Fig. 1-44 Peak detector (BB)

PEAK DETECTOR

The 4084/25 peak detector is a special type of sample/hold amplifier. The input signal is acquired and tracked (PEAK DETECT mode) until it reaches a maximum value, then the unit automatically holds this value while signaling that a peak has been reached (STATUS output). The 4084/25 can then be placed in the HOLD mode to ignore further peaks or RESET to a reference level, ready to detect the next peak. The extremely low output droop (voltage decay with time) of this unit allows it to be used with a variety of instruments to record or display its output (analog-to-digital converters, digital voltmeters, DPM's, analog meters, etc.).

The 4084/25 will detect peaks in the range of − 10 V to + 10 V. The RESET mode charges the internal holding capacitor to any reference level between + 10 V and − 10 V. The peak detector will then detect any peak more positive than the reference level. For instance, with a voltage reference input of 0 V, the unit will detect peak voltages between 0 V and + 10 V and with a − 10 V voltage reference input, the 4084/25 will detect peaks between − 10 V and + 10 V.

Contact Bounce Elimination

When the logic inputs are driven from relays and manual switches, care should be taken to eliminate contact bounce.

Contact bounce will cause the units to switch in and out of the PEAK-DETECT mode, each time adding several millivolts of charge offset. The circuit shown will eliminate contact bounce.

Using the 4084/25 With an A/D Converter

The 4084/25 may be very easily used with an analog-to-digital converter or a digital voltmeter. The STATUS output will start the conversion immediately after a peak has been detected. The A/D converter will start converting as soon as a peak voltage is reached.

Measurement of Time to Peak

Since the STATUS output switches to a logical "0" when a peak is detected, it can be used to control a counter measuring the time from start of the PEAK-DETECT mode to a peak.

The counter is set to zero by the logic inputs when the 4084/25 is in the RESET mode. It starts counting when the module goes into PEAK-DETECT. It stops counting when the peak is reached and the STATUS output goes low.

Peak-to-Peak Detector

The figure shows a connection to measure the peak-to-peak value of a signal that swings both positive and negative in amplitude. The inverting amplifier A1 is necessary since the Model 4084/25 detects positive peaks only. Amplifier A2 takes the negative sum of the two peak detector outputs, that is, the peak-to-peak value of the input.

4084/25

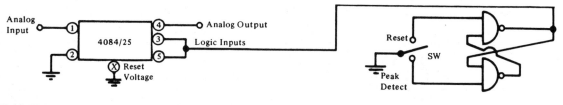

Fig. 1-45   Anti-bounce circuit.   (BB)

4084/25

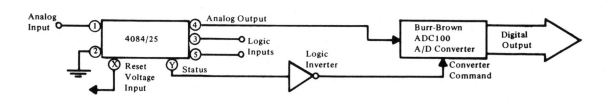

Fig. 1-46   Using the 4084/25 with an A/D converter.   (BB)

4084/25

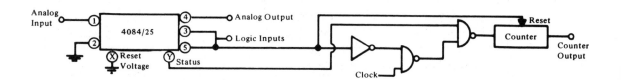

Fig. 1-47   Time to peak measurement.   (BB)

4084/25

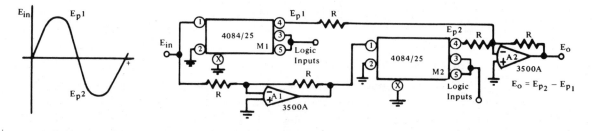

Fig. 1-48   Peak-to-peak detector   (BB)

4340

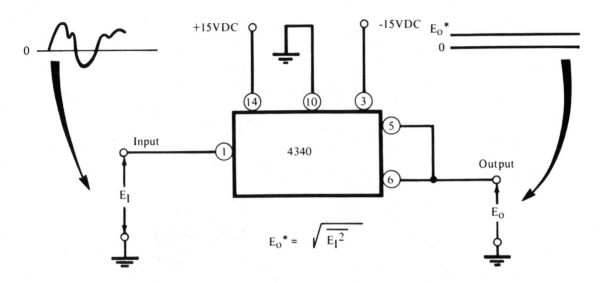

$$E_o{}^* = \sqrt{\overline{E_I{}^2}}$$

True RMS-to-dc Converter
Model 4340 RMS Converter - Connected to Produce Specified Unadjusted Accuracy.

Fig. 1-49   (BB)

4340   4340

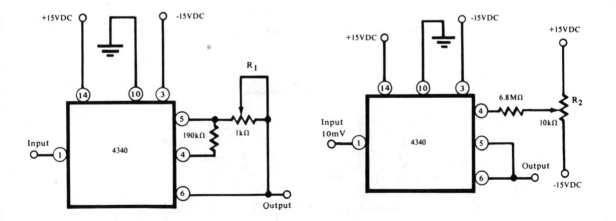

Fig. 1-50   Unity gain adjustment.   (BB)   Fig. 1-51   Offset voltage adjustment.   (BB)

4340   True RMS-to-dc converter

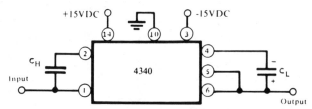

$C_H* = 22pF$ to 100pF and $C_l \geqslant 3.0\mu F$ for "adjusted" frequency response range.

Frequency response adjustments.

Fig. 1-52   (BB)

4340

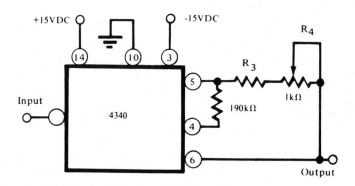

Set desired gain by selecting R_l such that $R_l = (A^2 - 1) \times 10k\Omega$. Apply appropriate mid-scale DC level to input and adjust R_4 for output equal to $A \times V_{INPUT}$ (VDC).

Non-unity gain adjustment.

Fig. 1-53   (BB)

4340

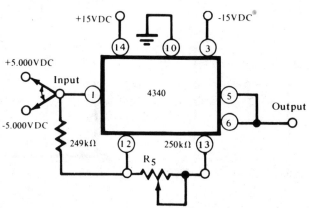

Alternately switch the input between +5.000VDC and -5.000VDC, adjust R_5 so that the output error voltage from +5.000VDC is the same for both input polarities.

Dc reversal error adjustment.

Fig. 1-54   (BB)

LOW COST TRUE RMS-TO-DC CONVERTER

If the 4341 is used in applications to measure complex waveforms, the following expanded trim procedure is recommended. First set all potentiometers at mid-turn position.

- ☐ Dc Reversal Error—Apply $+10.000$ V ±1 mV and -10.000 V ±1 mV to E_{in} alternatively, adjust R5 such that E_o readings are the same ±2 mV.
- ☐ Gain Adjustment—Apply $E_{in} = +10.000$ Vdc ±1 mV, adjust R1 such that E_o = $+10.000$ Vdc ±1 mV.
- ☐ Input Offset—Apply $+10.0$ mV ±0.1 mV and -10.0 mV ±0.1 mV to E_{in}, adjust R4 such that E_o readings are the same ±0.1 mV.
- ☐ Offset—Ground E_{in}, adjust R3 such that $E_o = 0 \pm0.1$ mV. Repeat previous step.
- ☐ Low Level Accuracy—Apply $E_{in} = +10.0$ mV ±1 mV, adjust R2 such that E_o = $+10.0$ mV ±0.1 mV.

Nonunity Gains

Gain values greater than unity can be achieved by inserting resistor R_x between pin 5 and pin 6. $R_x \simeq (A^2 - 1) \times 10\text{k} + 2\text{k}$ where A is the desired value of gain $(1 < A \leq 10)$. (R_x is in ohms).

4341

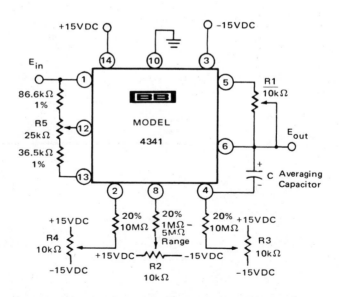

EXPANDED trim procedure (high accuracy applications).

Fig. 1-55

(BB)

4341

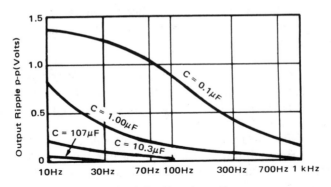

1.00V RMS Sine Wave Input Frequency

Fig. 1-56   Output ripple magnitude vs. input signal frequency.   (BB)

4341

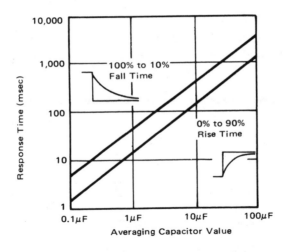

Averaging Capacitor Value

Fig. 1-57   Response time vs. value of averaging capacitor.   (BB)

VOLTAGE TO CURRENT POWER AMP

The speed, simplicity and efficiency available from switching current to voltage converters, make them useful for lower frequency ac waveform generation. With the input sampled at a 13 kHz rate, the input voltage is fed into the AD 460 VFC. The frequency outputted is changed into a train of precision pulses by the one shot, and inputted to the fast driver stage of the VN1210 output transistor.

The accuracy of this circuit is determined by the precision of the converter stages, and the ability of the output transistor to produce fast transition rise and fall pulses. This is accomplished by driving the VN1210 with the 3 transistor pre-driver stage capable of over 2 ampere source/sink currents, producing output current transition times of under 15 nanoseconds: allowing output frequency bandwidth to exceed 10 kHz. The sample/hold circuit is only necessary for bandwidths in excess of 100 Hz. Small size and exceptional efficiency, make this circuit attractive for portable, battery-powered equipment.

VN1210N5

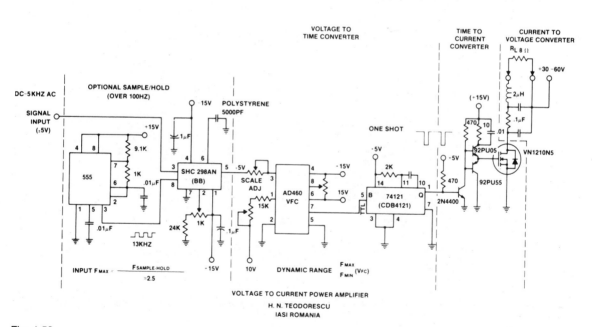

VOLTAGE TO CURRENT POWER AMPLIFIER

H. N. TEODORESCU
IASI ROMANIA

Fig. 1-58

(SU)

OPA11HT

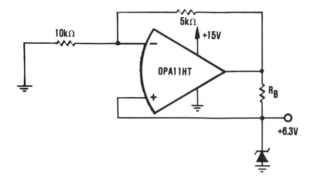

Fig. 1-59   Simple current source for improving the supply rejection of a zener reference voltage.   (BB)

TAA761C

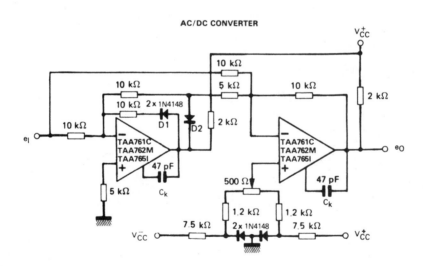

Fig. 1-60   (TH)

INA104

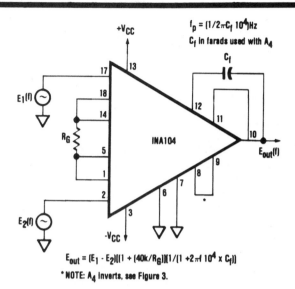

$f_p = (1/2\pi C_f \, 10^4)Hz$

C_f in farads used with A_4

$E_{out} = (E_1 - E_2)[(1 + (40k/R_G)][1/(1 + 2\pi f \, 10^4 \times C_f)]$

* NOTE: A_4 inverts, see Figure 3.

Fig. 1-61	Active low pass filtering.	(BB)

INA105

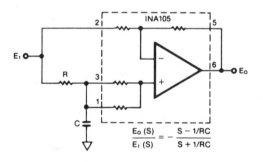

$$\frac{E_o(S)}{E_i(S)} = -\frac{S - 1/RC}{S + 1/RC}$$

$$\frac{E_o(S)}{E_i(S)} = +\frac{S - 1/RC}{S + 1/RC}$$

All-pass filter (provides unity gain and 0° to 180° phase shift output for frequencies of dc to ∞Hz).

All-pass filter (provides unity gain and −180° to 0° phase shift output for frequencies of dc to ∞Hz).

Fig. 1-62	For other INA105 circuits see Chapters 2 and 5.	(BB)

OPA2111

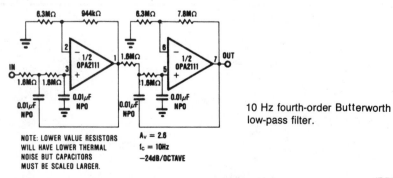

NOTE: LOWER VALUE RESISTORS
WILL HAVE LOWER THERMAL
NOISE BUT CAPACITORS
MUST BE SCALED LARGER.

$A_v = 2.6$
$f_c = 10Hz$
−24dB/OCTAVE

10 Hz fourth-order Butterworth low-pass filter.

Fig. 1-63		(BB)

OPA2111

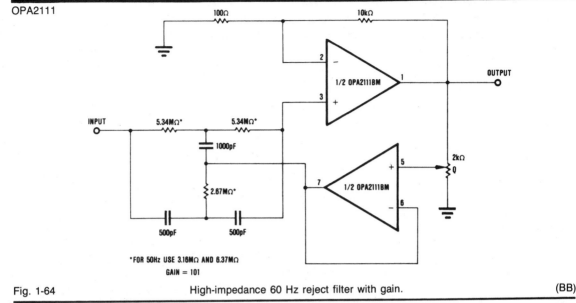

Fig. 1-64   High-impedance 60 Hz reject filter with gain.   (BB)

UAF11, UAF21   Universal Active Filter

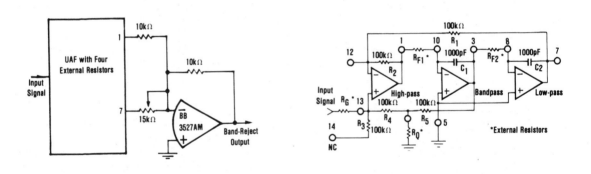

Fig. 1-65   Band-reject configuration.   Noninverting configuration.   (BB)

UAF11, UAF21

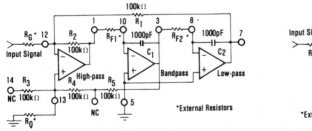

Fig. 1-66   Inverting configuration   Bi-quad configuration   (BB)

UAF11, UAF21

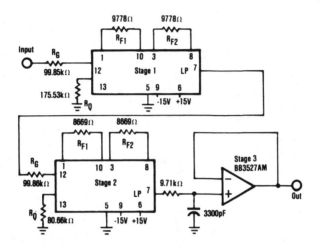

LM124

ACTIVE BANDPASS FILTER

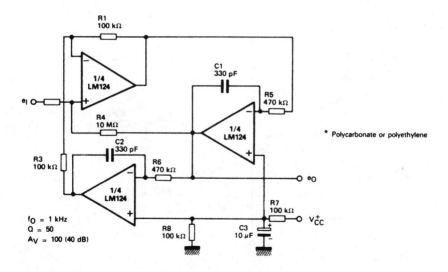

Fig. 1-68   (TH)

LM1458, TL071

TUNABLE NOTCH FILTER

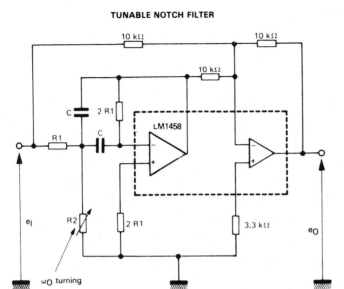

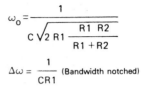

$$\omega_O = \cfrac{1}{C \sqrt{2\, R1\, \cfrac{R1\ R2}{R1 + R2}}}$$

$$\Delta\omega = \frac{1}{CR1} \quad \text{(Bandwidth notched)}$$

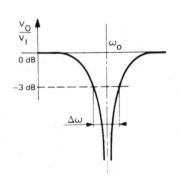

LOW-PASS FILTER

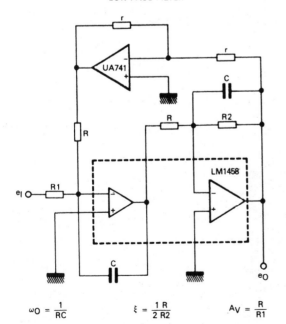

HIGH Q NOTCH FILTER

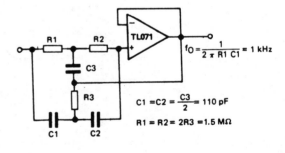

$$f_O = \frac{1}{2\,\pi\, R1\, C1} = 1\ \text{kHz}$$

$$C1 = C2 = \frac{C3}{2} = 110\ \text{pF}$$

$$R1 = R2 = 2R3 = 1.5\ \text{M}\Omega$$

$$\omega_O = \frac{1}{RC} \qquad \xi = \frac{1}{2}\frac{R}{R2} \qquad A_V = \frac{R}{R1}$$

Fig. 1-69

(TH)

TL074

POSITIVE FEEDBACK BANDPASS-FILTER

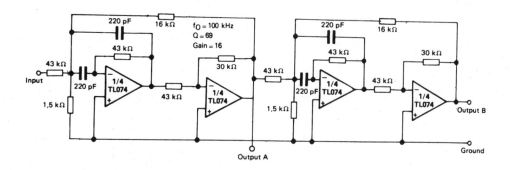

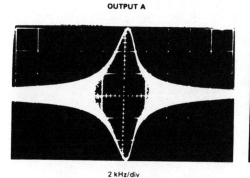

OUTPUT A

2 kHz/div

SECOND ORDER BANDPASS FILTER

$f_O = 100$ kHz ; $Q = 69$; Gain = 16

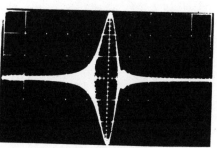

OUTPUT B

2 kHz/div

CASCADED BANDPASS FILTER

$f_O = 100$ kHz ; $Q = 30$; Gain = 4

Fig. 1-70

(TH)

2
Amplifier Circuits

3554

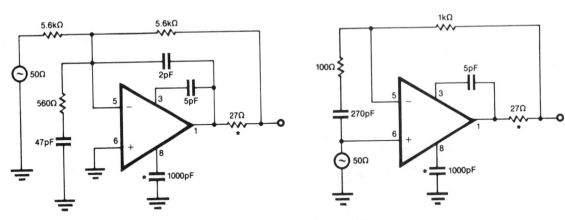

INVERTING AMPLIFIER

NONINVERTING BUFFER AMPLIFIER

*Needed only for driving capacitive loads

(BB)

Fig. 2-1

3554

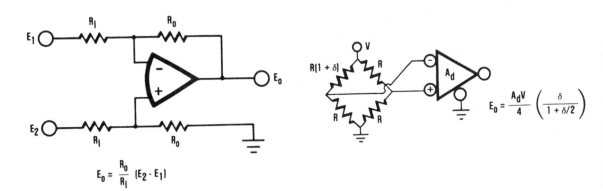

$$E_0 = \frac{R_0}{R_I} (E_2 - E_1)$$

$$E_0 = \frac{A_dV}{4} \left(\frac{\delta}{1 + \delta/2} \right)$$

Fig. 2-2   Single op-amp differential amplifier.   Bridge amplifier with one active bridge arm.   (BB)

OPA201

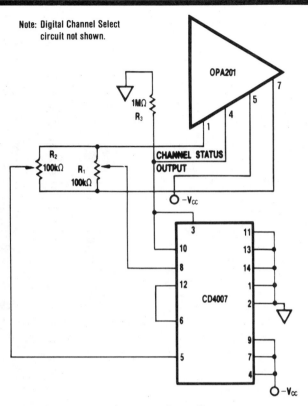

Note: Digital Channel Select
circuit not shown.

Independent dual-offset adjustment.

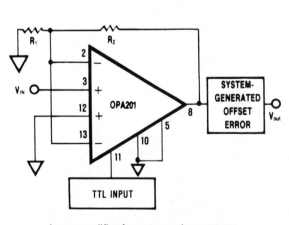

Input amplifier for auto-zeroing systems.

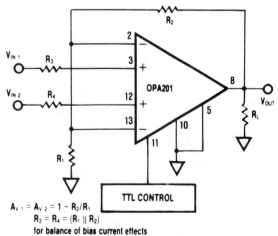

$A_{v\,1} = A_{v\,2} = 1 + R_2/R_1$
$R_3 = R_4 = (R_1 \parallel R_2)$
for balance of bias current effects
$[R_L \quad (R_2 + R_1)] \geq 10k\Omega$ for output current rating

Selectable input amplifier, noninverting.

Fig. 2-3

(BB)

OPA201

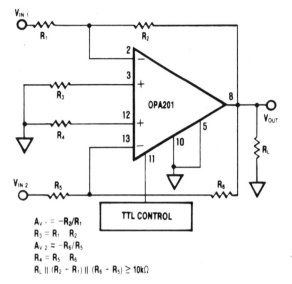

$A_{V\,1} = -R_2/R_1$
$R_3 = R_1\ \ R_2$
$A_{V\,2} = -R_6/R_5$
$R_4 = R_5\ \ R_6$
$R_L \parallel (R_2 - R_1) \parallel (R_6 - R_5) \geq 10k\Omega$

Selectable input amplifier, inverting.

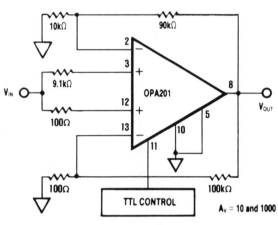

Switchable gain amplifier.

$A_V = 10 \text{ and } 1000$

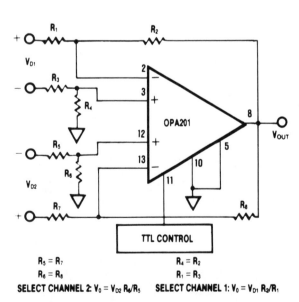

$R_5 = R_7$
$R_6 = R_8$
$R_4 = R_2$
$R_1 = R_3$

SELECT CHANNEL 2: $V_0 = V_{D2}\,R_6/R_5$ **SELECT CHANNEL 1:** $V_0 = V_{D1}\,R_2/R_1$

Low power dual-channel differential amplifier.

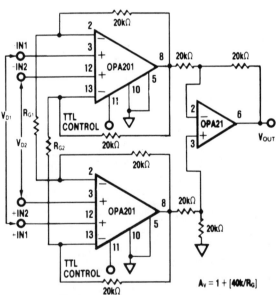

$A_V = 1 + [40k/R_G]$

Low power dual-channel instrumentation amplifier.

Fig. 2-3. Continued from page 57.

(BB)

FAST SUMMING AMPLIFIER WITH LOW INPUT CURRENT

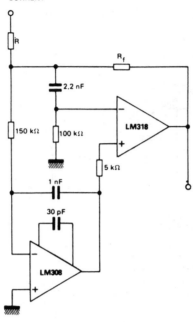

Fig. 2-4

(TH)

LM158, LM258, LM358, LM2904

HIGH INPUT IMPEDANCE, DC DIFFERENTIAL AMPLIFIER

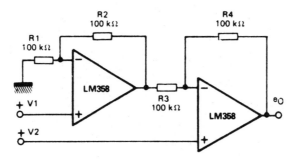

for $\dfrac{R1}{R2} = \dfrac{R4}{R3}$ (CMRR depends on this resistor ratio match)

$V_O = (1 + \dfrac{R4}{R3})\ (V2 - V1)$

As shown : $V_O = 2(V2 - V1)$

USING SYMMETRICAL AMPLIFIERS TO REDUCE INPUT CURRENT (GENERAL CONCEPT)

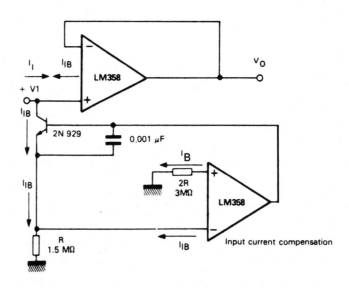

Fig. 2-5

LM158, LM258, LM358, LM2904

HIGH INPUT Z ADJUSTABLE-GAIN DC INSTRUMENTATION AMPLIFIER

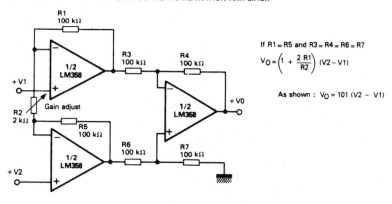

If R1 = R5 and R3 = R4 = R6 = R7

$$V_O = \left(1 + \frac{2\,R1}{R2}\right)(V2 - V1)$$

As shown : $V_O = 101\,(V2 - V1)$

Fig. 2-6

(TH)

LM124, LM224, LM324, LM2902

HIGH INPUT Z, DC DIFFERENTIAL AMPLIFIER

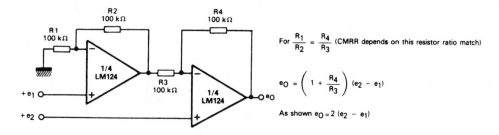

For $\dfrac{R_1}{R_2} = \dfrac{R_4}{R_3}$ (CMRR depends on this resistor ratio match)

$$e_O = \left(1 + \frac{R_4}{R_3}\right)(e_2 - e_1)$$

As shown $e_O = 2\,(e_2 - e_1)$

USING SYMMETRICAL AMPLIFIERS TO REDUCE INPUT
CURRENT (GENERAL CONCEPT)

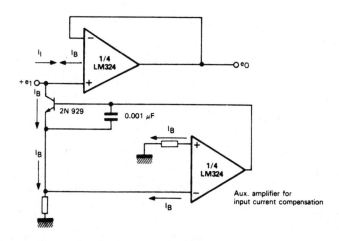

Fig. 2-7

LM124, LM224, LM324, LM2902

HIGH INPUT Z ADJUSTABLE GAIN DC INSTRUMENTATION AMPLIFIER

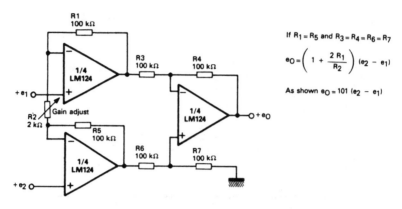

If $R_1 = R_5$ and $R_3 = R_4 = R_6 = R_7$

$$e_O = \left(1 + \frac{2 R_1}{R_2} \right) (e_2 - e_1)$$

As shown $e_O = 101 (e_2 - e_1)$

Fig. 2-8

(TH)

LM158, LM258, LM358, LM2904

AC COUPLED INVERTING AMPLIFIER

$$A_V = \frac{-R1}{R} \text{ (As shown } A_V = -10)$$

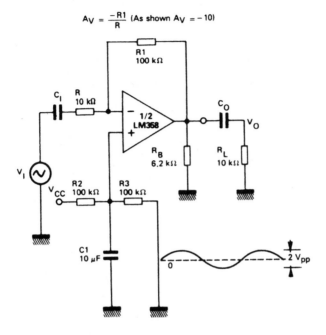

AC COUPLED NON INVERTING AMPLIFIER

$$\text{Gain} = 1 + \frac{R2}{R1} \text{ (As shown, Gain} = 11)$$

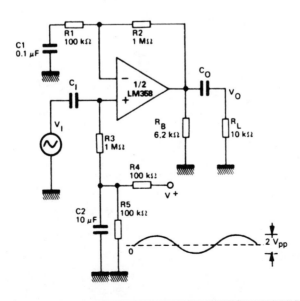

Fig. 2-9

(TH)

LM158, LM258, LM358, LM2904

NON-INVERTING DC AMPLIFIER

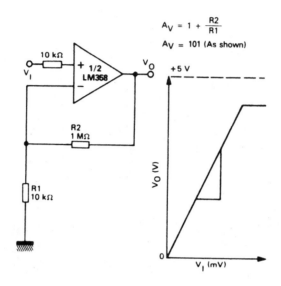

$$A_V = 1 + \frac{R2}{R1}$$

$$A_V = 101 \text{ (As shown)}$$

DC SUMMING AMPLIFIER

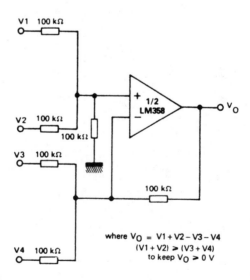

where $V_O = V1 + V2 - V3 - V4$
$(V1 + V2) \geqslant (V3 + V4)$
to keep $V_O \geqslant 0$ V

Fig. 2-10

(TH)

LM193

BASIC COMPARATOR

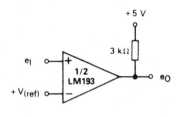

LOW FREQUENCY OP AMP

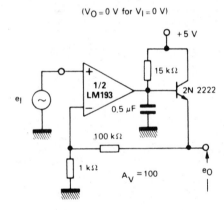

DRIVING CMOS

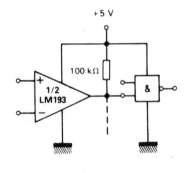

LOW FREQUENCY OP AMP

$(V_O = 0 \text{ V for } V_I = 0 \text{ V})$

DRIVING TTL

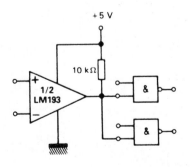

ZERO CROSSING DETECTOR

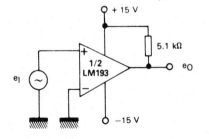

LM193

LOW FREQUENCY OP AMP WITH OFFSET ADJUST

ZERO CROSSING DETECTOR (SINGLE POWER SUPPLY)

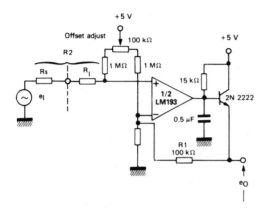

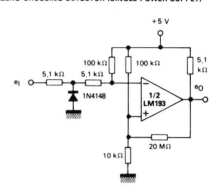

Fig. 2-12

(TH)

LM124

Op-amp circuits

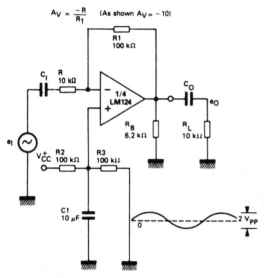

$A_V = \frac{-R}{R_1}$ (As shown $A_V = -10$)

AC COUPLED INVERTING AMPLIFIER

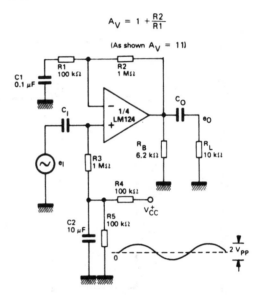

$A_V = 1 + \frac{R2}{R1}$

(As shown $A_V = 11$)

AC COUPLED NON-INVERTING AMPLIFIER

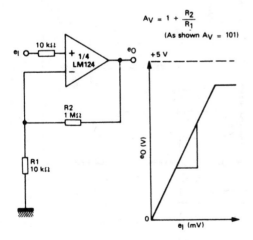

$A_V = 1 + \frac{R_2}{R_1}$

(As shown $A_V = 101$)

NON-INVERTING DC GAIN

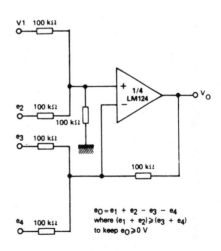

$e_O = e_1 + e_2 - e_3 - e_4$
where $(e_1 + e_2) \geqslant (e_3 + e_4)$
to keep $e_O \geqslant 0$ V

DC SUMMING AMPLIFIER

Fig. 2-13   (TH)

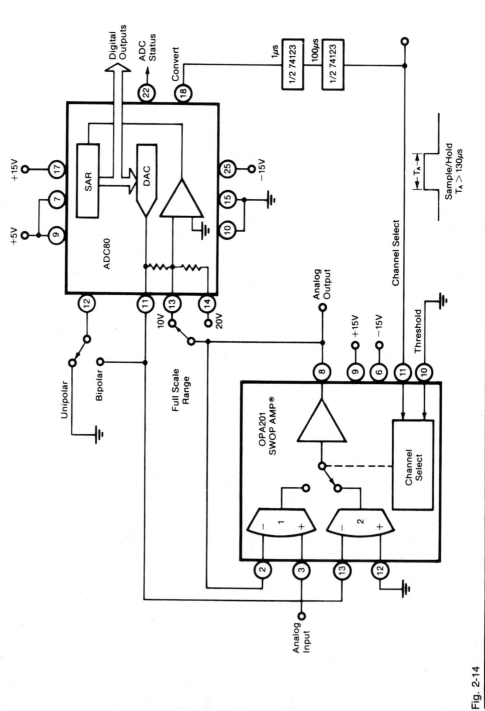

Switchable-input op amp, 12-bit A/D converter, two one-shot multivibrators used to configure a zero-droop sample/hold amplifier

OPA201

Fig. 2-14

(BB)

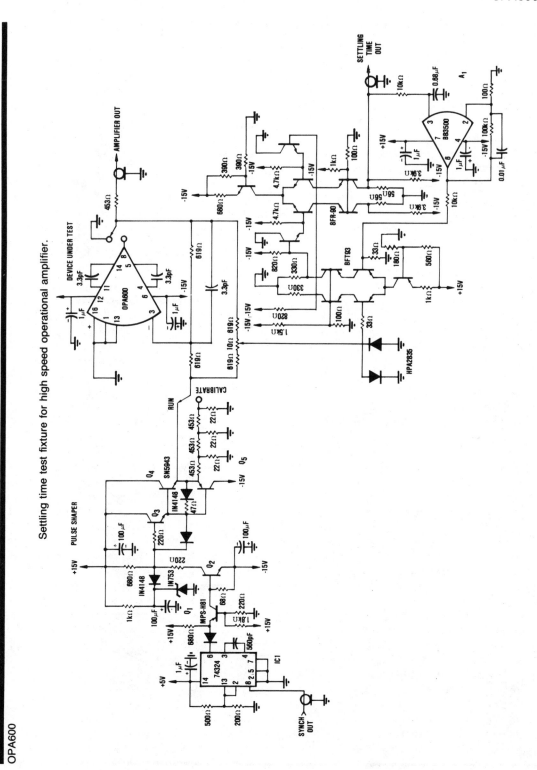

Settling time test fixture for high speed operational amplifier.

Fig. 2-15

(BB)

LM111, LM139

Additional op-amp circuits

STROBE

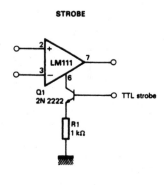

LIMIT COMPARATOR

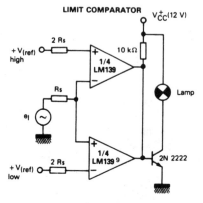

SPLIT-SUPPLY APPLICATIONS

INCREASING INPUT STAGE CURRENT

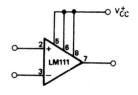

ZERO CROSSING DETECTOR

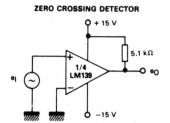

OFFSET BALANCING

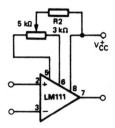

COMPARATOR WITH A NEGATIVE REFERENCE

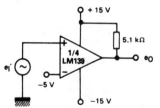

Fig. 2-16

(TH)

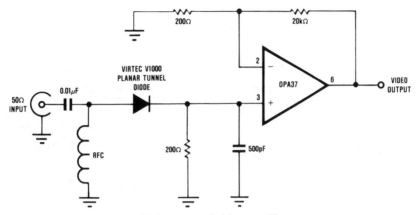

Rf detector and video amplifier.

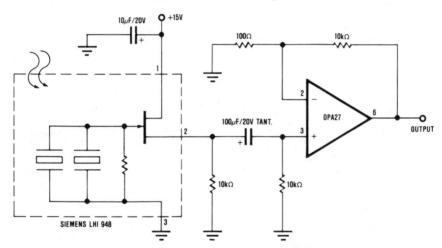

Balanced pyroelectric infrared detector.

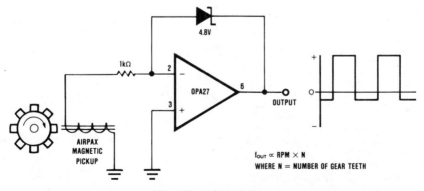

Magnetic tachometer.

Fig. 2-17

OPA27, OPA37

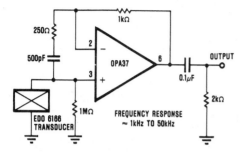

Hydrophone Preamplifier

GAIN = −1010V/V
$V_{os} \approx 2\mu V$
DRIFT $\approx 0.07\mu V/°C$
$e_n \approx 1nV/\sqrt{Hz}$ at 10Hz
 $0.9nV/\sqrt{Hz}$ at 100Hz
 $0.87nV/\sqrt{Hz}$ at 1kHz
FULL POWER BANDWIDTH $\approx$ 180kHz
GAIN BANDWIDTH $\approx$ 500MHz
EQUIVALENT NOISE RESISTANCE $\approx$ 50Ω

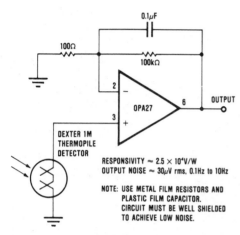

Long-wavelength infrared detector amplifier.

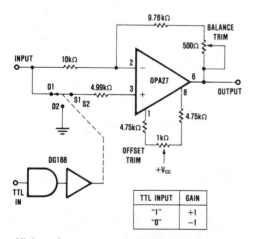

High performance synchronous demodulator.

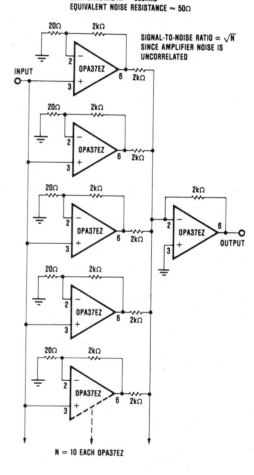

Ultra-low noise "N" stage parallel amplifier

Fig. 2-18

OPA27, OPA37

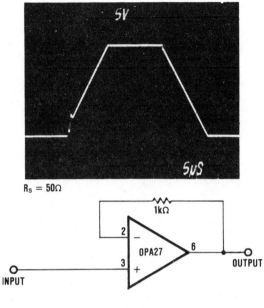

$R_S = 50\Omega$

Unity-gain buffer.

High slew rate unity-gain buffer.

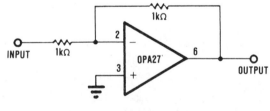

Unity-gain inverting amplifier.

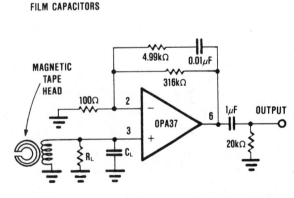

METAL FILM RESISTORS
FILM CAPACITORS

$G \approx 50dB$ at 1kHz

R_L AND C_L PER HEAD MANUFACTURER'S RECOMMENDATIONS

NAB tape head preamplifier.

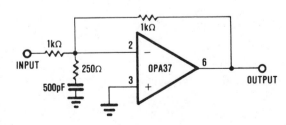

High slew rate unity-gain inverting amplifier.

Fig. 2-19

(BB)

OPA27, OPA37

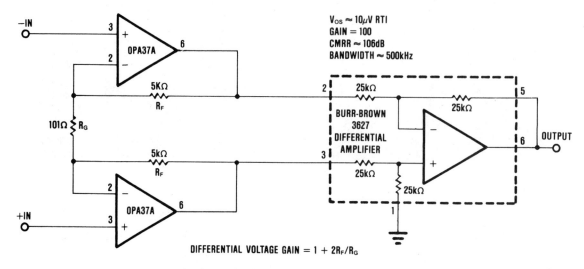

Low noise instrumentation amplifier.

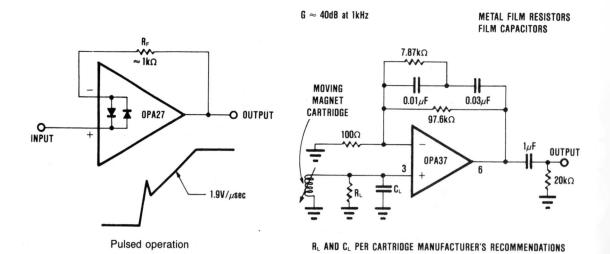

Pulsed operation

Low-noise RIAA preamplifier.

Fig. 2-19. Continued from page 73.

(BB)

OPA27, OPA37

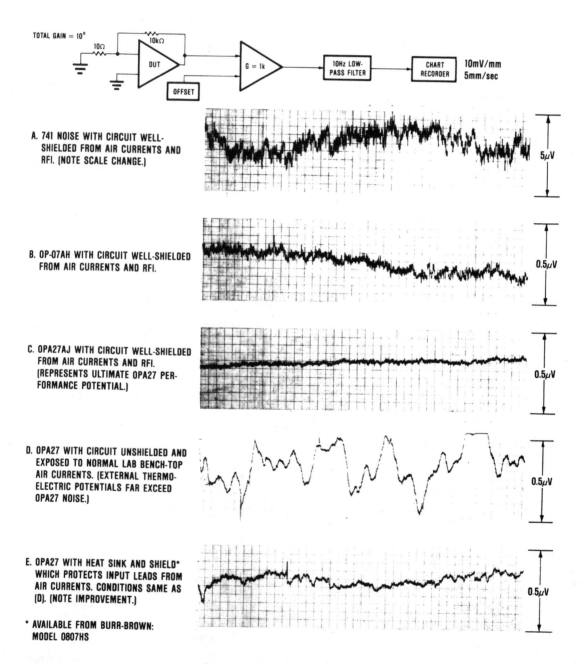

Low frequency noise comparison.

Fig. 2-20

OPA111

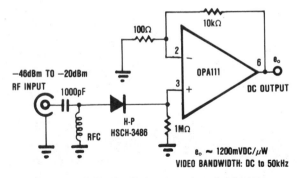

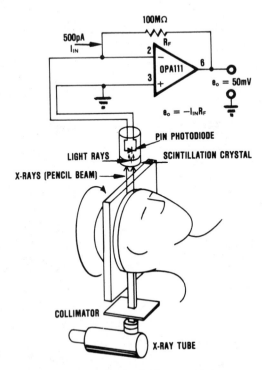

Zero-bias Schottky diode square-law rf detector.

$e_o \sim 1200mVDC/\mu W$
VIDEO BANDWIDTH: DC to 50kHz

Computerized axial tomography
(CAT) scanner channel amplifier.

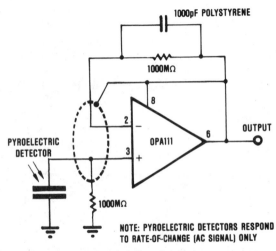

NOTE: PYROELECTRIC DETECTORS RESPOND
TO RATE-OF-CHANGE (AC SIGNAL) ONLY

Pyroelectric infrared detector.

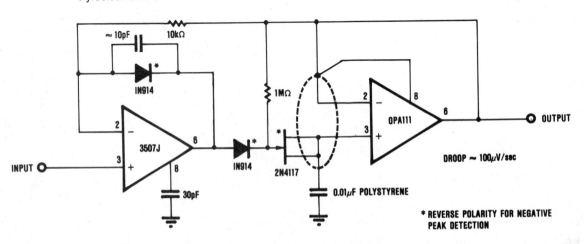

Fig. 2-21   Low-droop positive peak detector.   (BB)

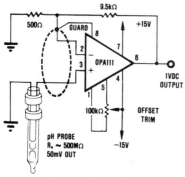

High impedance (10^{14} Ω) amplifier.

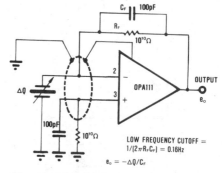

Piezoelectric transducer charge amplifier.

LOW FREQUENCY CUTOFF = $1/(2\pi R_F C_F) = 0.16$Hz

$e_o = -\Delta Q/C_F$

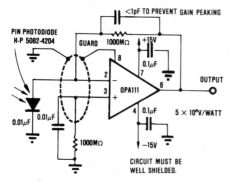

Sensitive photodiode amplifier.

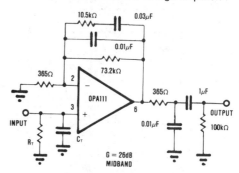

$G = 26$dB
MIDBAND

RIAA equalized phono preamplifier.

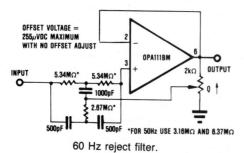

60 Hz reject filter.

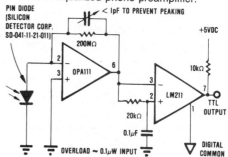

High sensitivity (under 1 nW) fiber
optic receiver for 9600 Baud Manchester data.

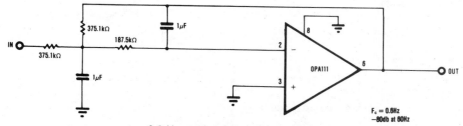

Fig. 2-22   0.6 Hz second order low-pass filter.   (BB)

OPA111

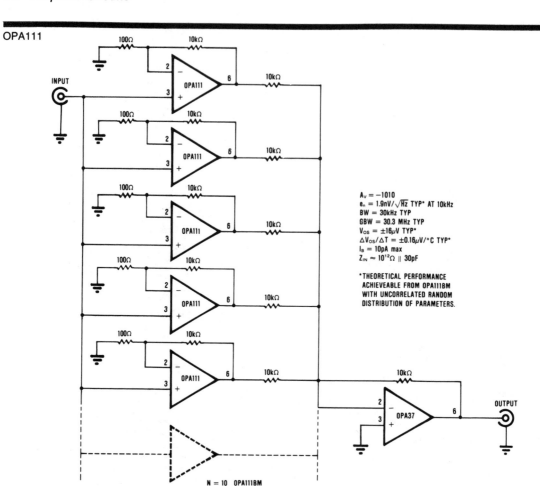

$A_V = -1010$
$e_n = 1.9nV/\sqrt{Hz}$ TYP* AT 10kHz
BW = 30kHz TYP
GBW = 30.3 MHz TYP
$V_{OS} = \pm16\mu V$ TYP*
$\Delta V_{OS}/\Delta T = \pm0.16\mu V/°C$ TYP*
$I_B = 10pA$ max
$Z_{IN} \approx 10^{12}\Omega \parallel 30pF$

*THEORETICAL PERFORMANCE
ACHIEVEABLE FROM OPA111BM
WITH UNCORRELATED RANDOM
DISTRIBUTION OF PARAMETERS.

N = 10 OPA111BM

'N' stage parallel-input amplifier for reduced relative amplifier noise at the output.

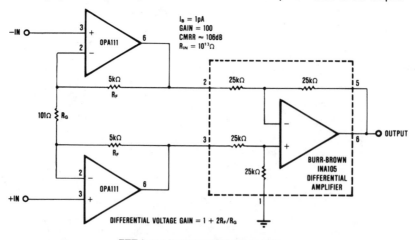

$I_B = 1pA$
GAIN = 100
CMRR ≈ 106dB
$R_{IN} = 10^{13}\Omega$

DIFFERENTIAL VOLTAGE GAIN = $1 + 2R_F/R_G$

BURR-BROWN
INA105
DIFFERENTIAL
AMPLIFIER

Fig. 2-23   FET input instrumentation amplifier.   (BB)

TAA761C, TL071, TL074

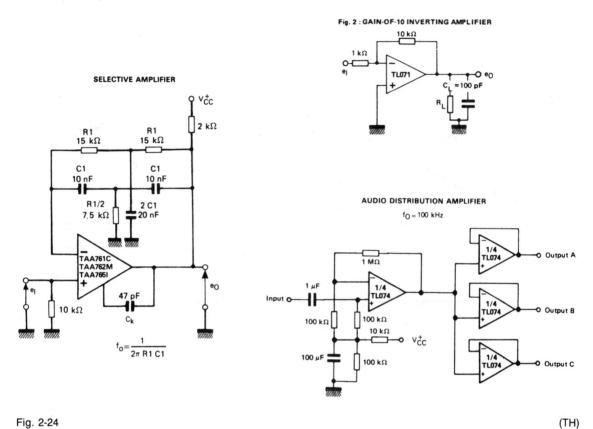

SELECTIVE AMPLIFIER

$$f_O = \frac{1}{2\pi\,R1\,C1}$$

Fig. 2 : GAIN-OF-10 INVERTING AMPLIFIER

AUDIO DISTRIBUTION AMPLIFIER

$f_O = 100$ kHz

Fig. 2-24 (TH)

LM193

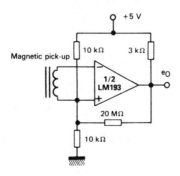

TRANSDUCER AMPLIFIER

Fig. 2-25 (TH)

INA104

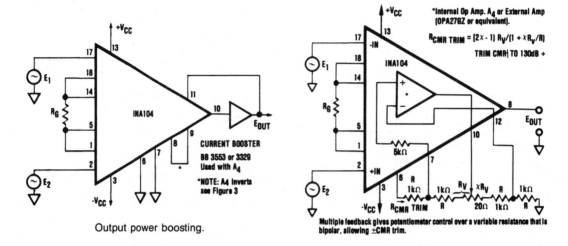

Output power boosting.

CMR trim.

Fig. 2-26

(BB)

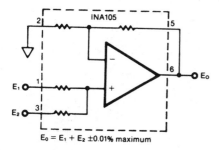

$E_o = E_1 + E_2 \pm 0.01\%$ maximum

Precision summing amplifier.

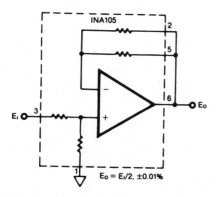

$E_o = E_1/2, \pm 0.01\%$

Precision (gain = ½) amplifier. Allows ±20 V input with ±15 V power supplies.

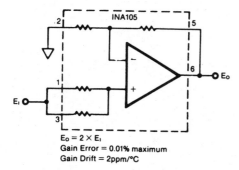

$E_o = 2 \times E_1$
Gain Error = 0.01% maximum
Gain Drift = 2ppm/°C

Precision (gain = 2) amplifier.

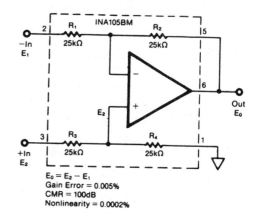

$E_o = E_2 - E_1$
Gain Error = 0.005%
CMR = 100dB
Nonlinearity = 0.0002%

Precision difference amplifier.

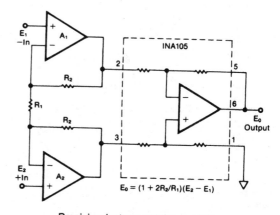

$E_o = (1 + 2R_2/R_1)(E_2 - E_1)$

Precision instrumentation amplifier.

For low source impedance applications, an input stage using OPA37 op amps will give the best low noise, offset, and temperature drift performance. At source impedances above about 10kΩ, the bias current noise of the OPA37 reacting with the input impedance begins to dominate the noise performance. For these applications, using the OPA111 or Dual OPA2111 FET input op amp will provide lower noise performance. For lower cost use the OPA121 plastic. To construct an electrometer use the OPA128.

A_1, A_2	R_1 (Ω)	R_2 (Ω)	Gain (V/V)	CMRR (dB)	Max I_B	Noise at 1kHz (nV/√Hz)
OPA37A	50.5	2.5k	100	128	40nA	4
OPA111B	202	10k	100	110	1pA	10
OPA128LM	202	10k	100	118	75fA	38

For other INA105 circuits see Chapters 1 and 5.

Fig. 2-27

3572

High-current, high power op-amp circuits

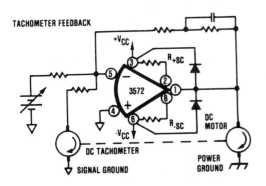

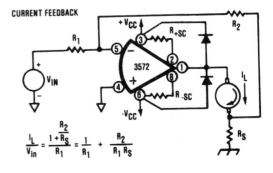

$$\frac{I_L}{V_{in}} = \frac{1 + \frac{R_2}{R_S}}{R_1} = \frac{1}{R_1} + \frac{R_2}{R_1 R_S}$$

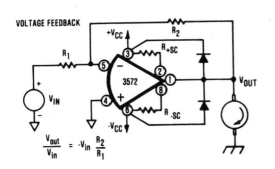

$$\frac{V_{out}}{V_{in}} = -V_{in} \frac{R_2}{R_1}$$

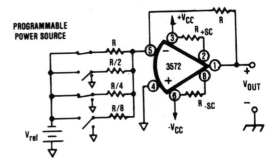

(BB)

Fig. 2-28

3584

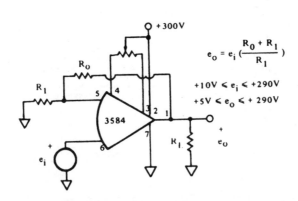

$$e_o = e_i \left(\frac{R_0 + R_1}{R_1} \right)$$

$+10V \leqslant e_i \leqslant +290V$

$+5V \leqslant e_o \leqslant +290V$

Another 3584 circuit is shown in Chapter 4.

(BB)

Fig. 2-29

OPA156A

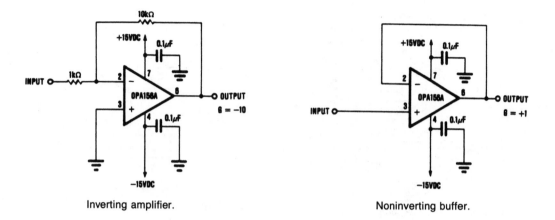

Inverting amplifier.   Noninverting buffer.

Fig. 2-30

(BB)

OPA201

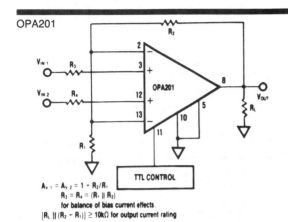

$A_{v1} = A_{v2} = 1 + R_2/R_1$
$R_3 = R_4 = (R_1 \parallel R_2)$
for balance of bias current effects
$[R_L \parallel (R_2 + R_1)] \geq 10k\Omega$ for output current rating

Selectable input amplifier, noninverting.

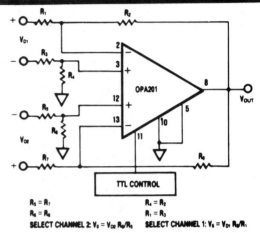

$R_5 = R_7$
$R_6 = R_8$
$R_4 = R_2$
$R_1 = R_3$

SELECT CHANNEL 2: $V_0 = V_{D2} R_6/R_5$ SELECT CHANNEL 1: $V_0 = V_{D1} R_6/R_1$

Low power dual-channel differential amplifier.

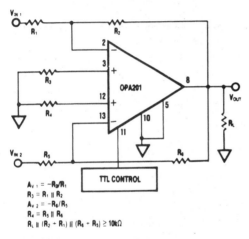

$A_{v1} = -R_2/R_1$
$R_3 = R_1 \parallel R_2$
$A_{v2} = -R_6/R_5$
$R_4 = R_5 \parallel R_6$
$R_L \parallel (R_2 + R_1) \parallel (R_6 + R_5) \geq 10k\Omega$

Selectable input amplifier, inverting.

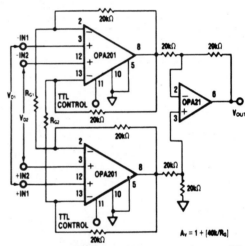

$A_v = 1 + [40k/R_G]$

Low power dual-channel instrumentation amplifier.

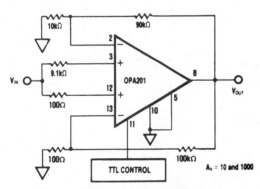

$A_v = 10$ and 1000

Fig. 2-31   Switchable gain amplifier.

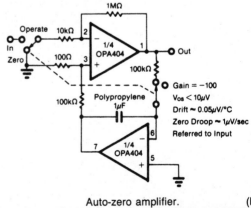

Gain = −100
$V_{OS} < 10\mu V$
Drift ≈ 0.05μV/°C
Zero Droop ≈ 1μV/sec
Referred to Input

Auto-zero amplifier.   (BB)

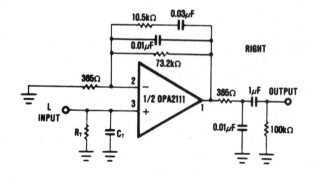

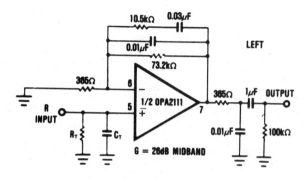

RIAA equalized stereo preamplifier.

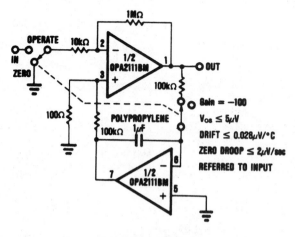

Auto-zero amplifier.

Fig. 2-32

3554   Op-amp circuits

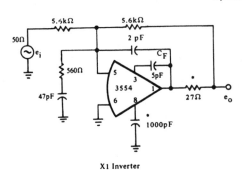

X1 Inverter

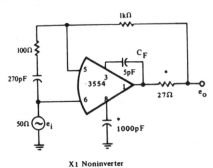

X1 Noninverter

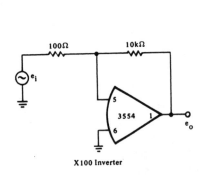

X10 Inverter

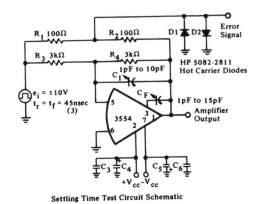

Settling Time Test Circuit Schematic

View from Component Side.
Shaded area is the pattern side conductor.

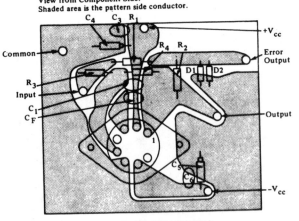

Settling Time Test Circuit Layout

X100 Inverter

NOTES:
1. These circuits are optimized for driving large capacitive loads (to 470pF).
2. The 3554 is stable at gains of greater than 55 ($C_L \leq 100pF$) without any frequency compensation.
3. 45nsec is optimum. Very fast rise times (10-20nsec) may saturate the input stage causing less than optimum settling time performance.
*Indicates component that may be eliminated when large capacitive loads are not being driven by the device.

Fig. 2-33

(BB)

OPA128

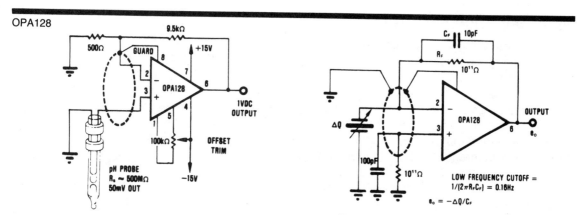

High impedance (10^{15} Ω) amplifier.

Piezoelectric transducer charge amplifier.

$$\text{LOW FREQUENCY CUTOFF} = 1/(2\pi R_F C_F) = 0.16 \text{Hz}$$

$$e_o = -\Delta Q/C_F$$

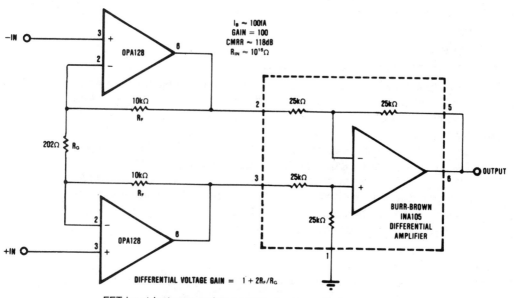

$I_B \sim 100 fA$
GAIN = 100
CMRR $\sim$ 118dB
$R_{IN} \sim 10^{15}$ Ω

BURR-BROWN
INA105
DIFFERENTIAL
AMPLIFIER

DIFFERENTIAL VOLTAGE GAIN = $1 + 2R_F/R_G$

FET input instrumentation amplifier for biomedical applications.

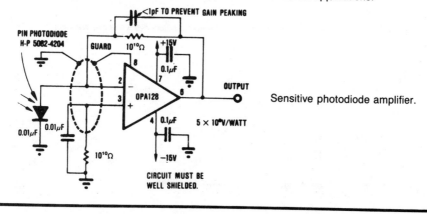

Sensitive photodiode amplifier.

5×10^6 V/WATT

CIRCUIT MUST BE
WELL SHIELDED.

Fig. 2-34

(BB)

INSTRUMENTATION AMPLIFIERS

An instrumentation amplifier is a closed-loop, differential input gain block. It is a committed circuit with the primary function of accurately amplifying the voltage applied to its inputs.

Ideally, the instrumentation amplifier responds only to the difference between the two input signals and exhibits extremely-high impedances between the two input terminals, and from each terminal to ground. The output voltage is developed single-ended with respect to ground and is equal to the product of amplifier gain and the difference of the two input voltages (see Fig. 2-35A).

The amplifier gain G is normally set by the user with a single external resistor. The properties of this model may be summarized as infinite input impedance, zero output impedance, the output voltage proportional to only the difference voltage $(e_2 - e_1)$, a precisely known gain constant (implying no nonlinearity), and unlimited bandwidth. This amplifier would completely reject signal components common to both inputs (common-mode rejection) and would exhibit no dc offset voltage or drift.

Characteristics of Instrumentation Amplifiers

It is desirable to achieve, as close as possible, the characteristics of the ideal instrumentation amplifier. The following paragraphs are a discussion of the, other-than-ideal, characteristics of the instrumentation amplifiers.

Input Impedance. A simple model of realistic instrumentation amplifier is shown in Fig. 2-35B. The impedance Z_{id} represents the differential input impedance. The common-mode input impedance Z_{icm} is represented as two equal components, $2Z_{icm}$, from each input to ground. These finite resistances contribute an effective gain error due to loading of the source resistance. The instrumentation amplifier provides a load on the source of $Z_i = Z_{id} \parallel Z_{icm}$. If source impedance is $R_S = R_{s1} + R_{s2}$, the gain error caused by this loading is:

$$\text{Gain Error} = 1 - \frac{Z_i}{Z_i + R_s} = \frac{R_s}{Z_i + R_s} \cong \frac{R_s}{Z_i} \text{ if } Z_i > R_s$$

Instrumentation amplifiers

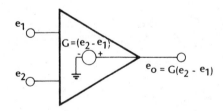

Idealized Model of an Instrumentation Amplifier

Fig. 2-35A   (BB)

Instrumentation Amplifier

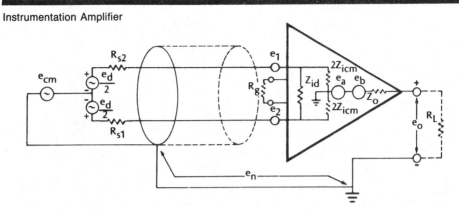

Simple Model of an instrumentation amplifier shown in a typical application configuration.

Fig. 2-35B   (BB)

If R_s is 10 kΩ and Z_i is 10 MΩ,

$$\text{Gain Error} \cong \frac{10 \times 10^3}{10 \times 10^6} = 0.1\%$$

The dc common-mode input impedance Z_{icm} will be independent of gain. The dc differential input impedance Z_{id} may vary as a function of gain. Specifications give the worst-case value. The nonzero output impedance of the amplifier will also create a gain error, the value of which depends on the load resistance.

Nonlinearity. The linearity of gain is possibly of more importance than the gain accuracy, since the value of the gain can be adjusted to compensate for simple gain errors. The nonlinearity is specified to be the peak deviation from a "best fit" straightline, expressed as a percent of peak-to-peak full scale output.

Common-mode Rejection. As illustrated in Fig. 2-35B, the output voltage has two components. One component is proportional to the differential input voltage $e_d = (e_2 - e_1)$. The second component is proportional to the common-mode input voltage. The common-mode voltage which appears at the amplifier's input terminals is defined as $E_{cm} = e_2 + e_1/2$. This may consist of some common-mode voltage in the source itself, e_{cm}, (such as bridge excitation) plus any noise voltage, e_n, between the source common and the amplifier common. As shown in Fig. 2-35B, the constant G represents the differential amplifier gain factor (fixed by the external gain-setting resistor). The constant (G/CMRR) represents the common-mode signal gain of the amplifier. The CMRR (common-mode rejection ratio) is the ratio of differential gain to common-mode gain. Thus CMRR is proportional to the differential gain and CMRR increases as the differential (gain G) increased. Hence, CMRR is usually specified for the maximum and the minimum values of gain of the amplifier. The common-mode rejection may be expressed in dB as - CMRR (dB) = $20 \log_{10}$ CMRR.

For an ideal instrumentation amplifier the output voltage component due to common-mode voltage should be zero. For a realistic instrumentation amplifier, the CMRR though very high, is still not infinite and so will cause an error voltage of E_{cm}/CMRR × G to appear at the output.

Source Impedance Unbalance. If the source impedances are unbalanced the source voltages ($e_{cm} + e_n$) are divided unequally upon the common-mode impedances and a differential signal is developed at the amplifier's input. This error signal cannot be separated from the desired signal. In the circuit in Figure 2 if $R_{s2} = O$, $R_{s1} = 1$ kΩ, $e_{cm} + e_n = 10$ V, and $Z_{cm} = 100$ MΩ, then the effect of unbalance is to generate a voltage.

$$e_2 - e_1 = 10 \text{ V} - 10 \text{ V} - \frac{10^8}{10^8 + 10^3} = 10 \text{ V} \frac{10^3}{10^8 + 10^3} \cong \frac{10 \text{ V}}{10^5} = 0.1 \text{mV}$$

If e_d full scale is 10 mV then this error is:

$$\text{Error} = \frac{0.1 \text{ mV}}{10 \text{ mV}} = 1\% \text{ of full scale.}$$

Offset Voltage and Drift. Most instrumentation amplifiers are two stage devices—they have a variable gain input stage and a fixed gain output stage. If V_i and V_o are the offset voltages of the input and output stages respectively, then the amplifiers total offset voltage referred to the input (RTI) = $V_i + V_o/G$ where G is the amplifier's gain. [Note that E_{OS} (RTI) × G.]

The initial offset voltage is usually adjustable to zero and therefore, the voltage drift is the more significant term since it cannot be nulled. The offset voltage drift also has two components—one due to the input stage of the amplifier and the other due to the output stage.

When the amplifier is operated at high gain, the drift of the input stage predominates. At low values of gain, the drift of the output stage will be the major component of drift. When the total output drift is referred to the input, the effective input voltage drift is largest for low values of gain. Output voltage drift will always be lowest at low gains. If $\Delta V_i/\Delta T = 2 \mu V/° C$ and $\Delta V_o/\Delta T = 500 \mu V/° C$ and the amplifier in a gain of 1000 V/V is nulled at 25° C, then at 65° C the offset voltage will be:

$$E_{os} \text{ (RTI) }_{65°}$$
$$= 40° \text{ C } [2 \mu V/° C + (500 \mu V/° C/1000 \text{ V/V})]$$
$$= 40° \text{ C } (2.5 \mu V/° C = 100 \mu V = 0.1 \text{ mV}$$

If the full scale input is 10 mV then the error due to voltage drift is:

$$\text{Error} = 0.1 \text{ mV/10 mV} = 1\% \text{ of full scale.}$$

Input Bias and Offset Currents. The input bias currents are the currents that flow out of (or into) either of the two inputs of the amplifier. They are the base currents for bipolar input stages and the JFET leakage currents for FET input stage. Offset currents are the difference of the two bias currents.

The bias currents flowing into the source resistances will generate offset voltages of $E_{os2} = 1_{B2} \times R_{s2}$ and $E_{os1} = 1_{B1} \times R_{s1}$. If $R_{s1} = R_{s2} = R_s/2$ the offset voltage at the input is $E_{os2} - E_{os1} = I_{os} \times R_x/2$. This input referred offset error may be compared directly with the input voltage to compute percent error. (Note that the source must be returned to power supply common or R_s will be infinite and the amplifier will saturate.)

OPA156A

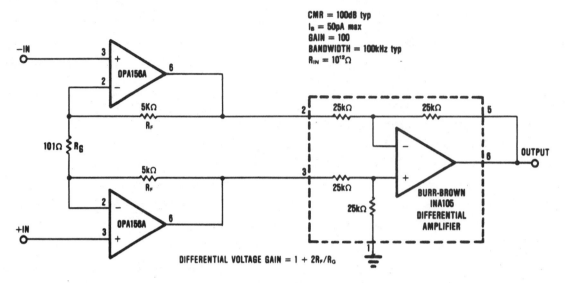

CMR = 100dB typ
I_B = 50pA max
GAIN = 100
BANDWIDTH = 100kHz typ
R_{IN} = $10^{12}\Omega$

DIFFERENTIAL VOLTAGE GAIN = 1 + 2R_F/R_G

Wideband FET input instrumentation amplifier.

Fig. 2-36   (BB)

INA110

Fast-settling FET-input very high accuracy instrumentation amplifier

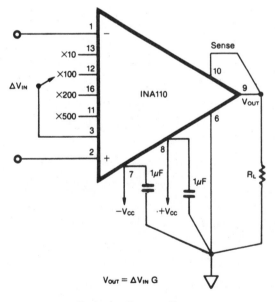

$V_{OUT} = \Delta V_{IN}\, G$

Basic circuit connection.

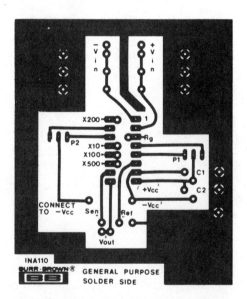

Recommended PC board payout for INA110.

Fig. 2-37   (BB)

INA110

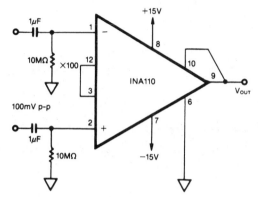

Ac-coupled differential amplifier for frequencies greater than 0.016 Hz.

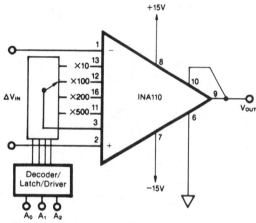

* Use manual switch or low resistance relay.
  Layout is critical (see section on Dynamic Performance).

Programmable-gain instrumentation amplifier (precision noninverting or inverting buffer with gain).

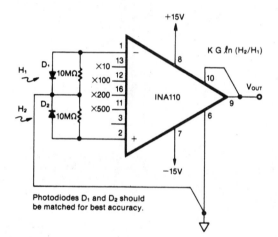

Photodiodes D_1 and D_2 should be matched for best accuracy.

Ratiometric light amplifier (absorbance measurement).

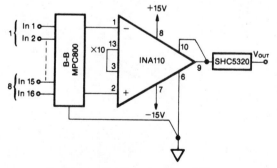

Rapid-scanning-rate data acquisition channel with 5 μs settling to 0.01%.

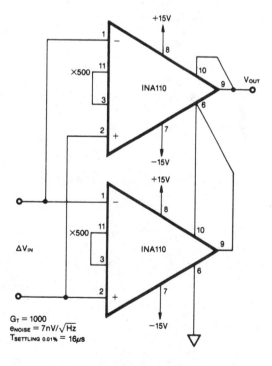

$G_T = 1000$
$e_{NOISE} = 7nV/\sqrt{Hz}$
$T_{SETTLING\ 0.01\%} = 16\mu s$

Fast-settling low-noise instrumentation amplifier with gain of 1000.

Fig. 2-38

(BB)

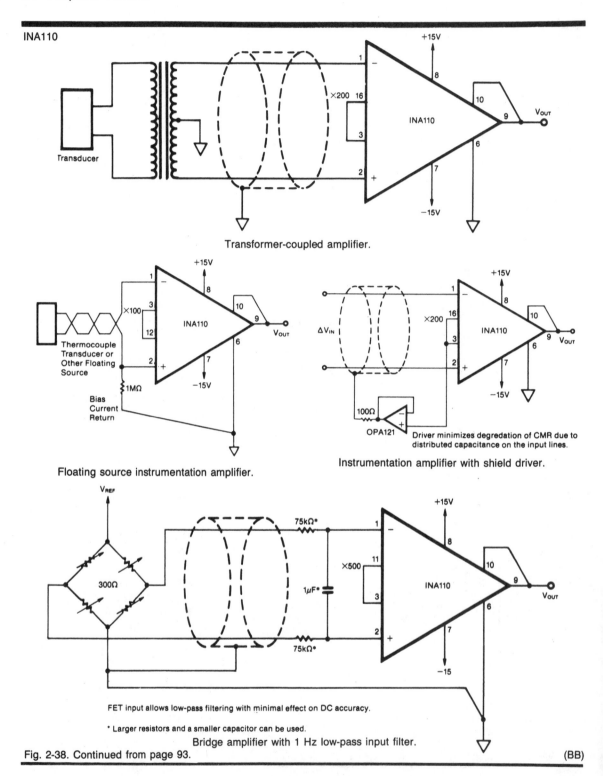

INA110

Transformer-coupled amplifier.

Floating source instrumentation amplifier.

Instrumentation amplifier with shield driver.

Driver minimizes degredation of CMR due to distributed capacitance on the input lines.

FET input allows low-pass filtering with minimal effect on DC accuracy.

* Larger resistors and a smaller capacitor can be used.

Bridge amplifier with 1 Hz low-pass input filter.

Fig. 2-38. Continued from page 93.

(BB)

INA110

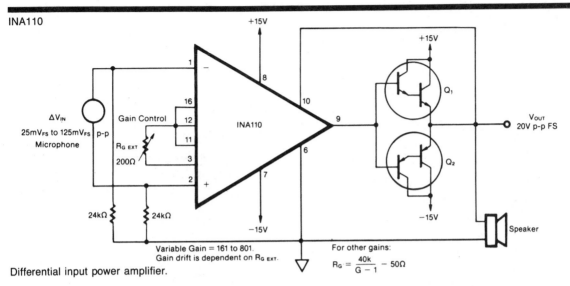

Differential input power amplifier.

Variable Gain = 161 to 801.
Gain drift is dependent on $R_{G\ EXT}$.

For other gains:
$$R_G = \frac{40k}{G-1} - 50\Omega$$

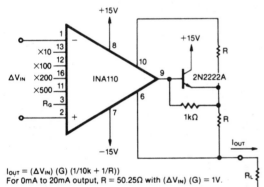

$$I_{OUT} = (\Delta V_{IN})(G)(1/10k + 1/R))$$
For 0mA to 20mA output, R = 50.25Ω with $(\Delta V_{IN})(G)$ = 1.

Differential input FET buffered current source.

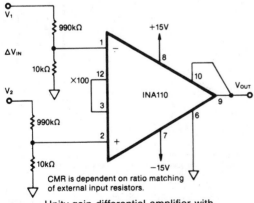

CMR is dependent on ratio matching
of external input resistors.

Unity-gain differential amplifier with
common-mode voltage range of 1000 V.

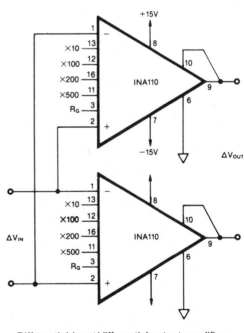

Differential input/differential output amplifier.

Fig. 2-39

(BB)

INA110

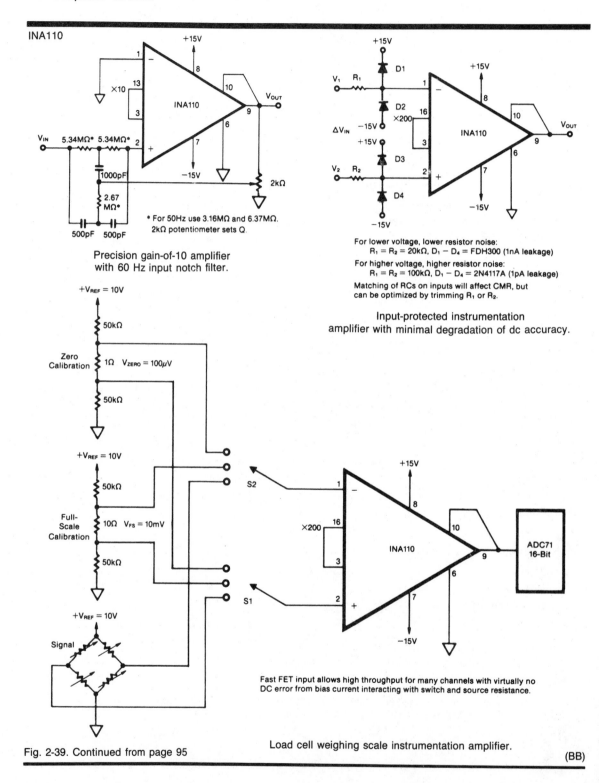

Precision gain-of-10 amplifier
with 60 Hz input notch filter.

* For 50Hz use 3.16MΩ and 6.37MΩ.
2kΩ potentiometer sets Q.

For lower voltage, lower resistor noise:
R₁ = R₂ = 20kΩ, D₁ − D₄ = FDH300 (1nA leakage)

For higher voltage, higher resistor noise:
R₁ = R₂ = 100kΩ, D₁ − D₄ = 2N4117A (1pA leakage)

Matching of RCs on inputs will affect CMR, but
can be optimized by trimming R₁ or R₂.

Input-protected instrumentation
amplifier with minimal degradation of dc accuracy.

Fast FET input allows high throughput for many channels with virtually no
DC error from bias current interacting with switch and source resistance.

Fig. 2-39. Continued from page 95

Load cell weighing scale instrumentation amplifier.

(BB)

INA110

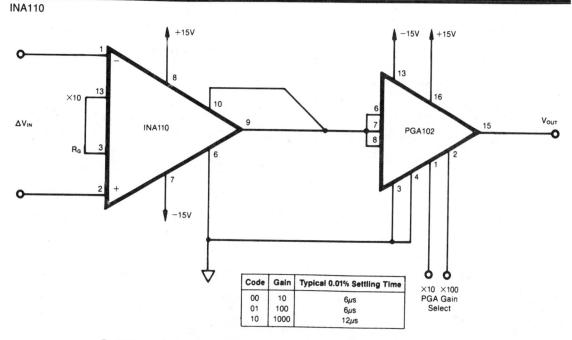

Code	Gain	Typical 0.01% Settling Time
00	10	6μs
01	100	6μs
10	1000	12μs

Digitially-controlled fast-settling programmable-gain instrumentation amplifier.

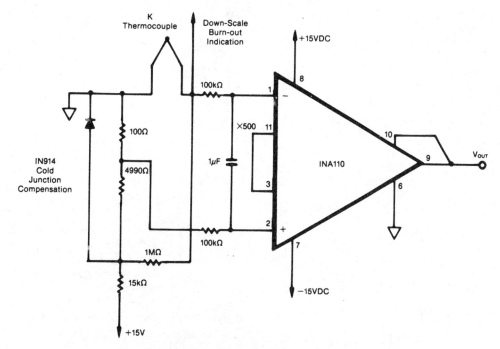

Fig. 2-40   Thermocouple amplifier with cold junction compensation and input low-pass filtering (<1 Hz).   (BB)

LM101A, LM201A, LM301A

INSTRUMENTATION AMPLIFIER

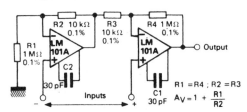

FAST INVERTING AMPLIFIER WITH
HIGH INPUT IMPEDANCE

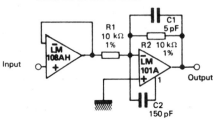

Fig. 2-41   (TH)

INA101   INA101

Instrumentation amplifiers

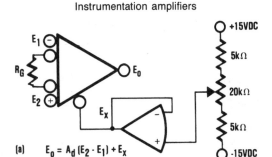

(a) $E_0 = A_d (E_2 - E_1) + E_x$

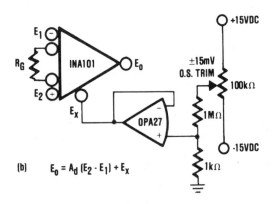

(b) $E_0 = A_d (E_2 - E_1) + E_x$

(a). Active output offset adjustment for modular
instrumentation amplifiers.

(b). Output offset adjustment of IA which is gain
independent.

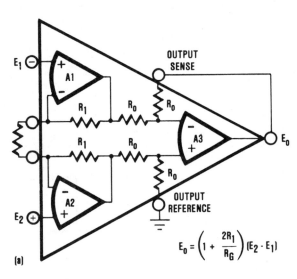

$$E_0 = \left(1 + \frac{2R_1}{R_G}\right)(E_2 - E_1)$$

(a)

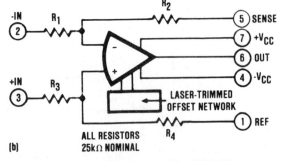

(b)

(a). Typical instrumentation amplifier (INA101).

(b). Low cost, unity-gain instrumentation amplifier (3627).

Fig. 2-42   (BB)   Fig. 2-43   (BB)

INA101

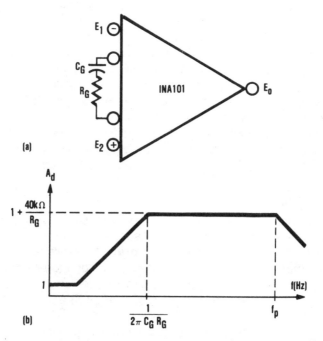

(a)

(b)

$$1 + \frac{40k\Omega}{R_G}$$

$$\frac{1}{2\pi C_G R_G}$$

IC instrumentation amplifier as a differential input high-pass filter.

Fig. 2-44

(BB)

3656

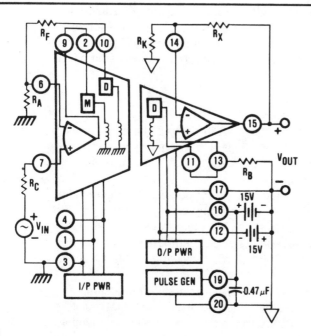

Burr-Brown 3656 as a high performance instrumentation amplifier with a CMR of 160 dB over a 3.5 kV common-mode range.

Fig. 2-45

(BB)

LOW POWER, HIGH ACCURACY, INSTRUMENTATION AMPLIFIER

The INA102 is a high-accuracy monolithic instrumentation amplifier designed for signal conditioning applications where low quiescent power is desired. On-chip thin-film resistors provide excellent temperature and stability performance. State-of-the-art laser trimming technology ensures high gain accuracy and common-mode rejection while avoiding expensive external components. These features make the INA102 ideally suited for battery powered and high volume applications.

The INA102 is also convenient to use. A gain of 1, 10, 100, or 1000 may be selected by simply strapping the appropriate pins together. 5 ppm/° C gain drift in low gains can then be achieved without external adjustment. When higher than specified CMR is required, CMR can be trimmed using the pins provided. In addition, balanced filtering can be accomplished in the output stage.

Discussion of Performance

Instrumentation amplifiers are differential input closed-loop gain blocks whose committed circuit accurately amplifies the voltage applied to their inputs. They respond mainly to the difference between the two input signals and exhibit extremely-high input impedance, both differential and common-mode. The feedback networks of this instrumentation amplifier are included on the monolithic chip. No external resistors are required for gains of 1, 10, 100 and 1000 in the INA102.

An operational amplifier, on the other hand, is an open-loop, uncommited device that requires external networks to close the loop. While op amps can be used to achieve the same basic function as instrumentation amplifiers, it is very difficult to reach the same level of performance. Using op amps often leads to design trade-offs when it is necessary to amplify low level signals in the presence of common-mode voltages while maintaining high input impedances. Figure 2-50A shows a simplified model of an instrumentation amplifier that eliminates most of the problems.

The INA102

A simplified schematic of the INA102 is shown. A three-amplifier configuration is used to provide the desirable characteristics of a premium performance instrumentation amplifier. In addition, it has features not normally found in integrated circuit instrumentation amplifiers.

The input buffers (A1 and A2) incorporate high performance, low drift amplifier circuitry. The amplifiers are connected in the noninverting configuration to provide the high input impedance (10^{10} Ω) desirable in instrumentation amplifier applications. The offset voltage and offset voltage versus temperature are low due to the monolithic design, and improved even further by state-of-the-art laser-trimming techniques.

The output stage (A3) is connected in a unity-gain differential amplifier configuration. A critical part of this stage is the matching of the four 20 kΩ resistors which provide the difference function. These resistors must be initially well matched

INA102

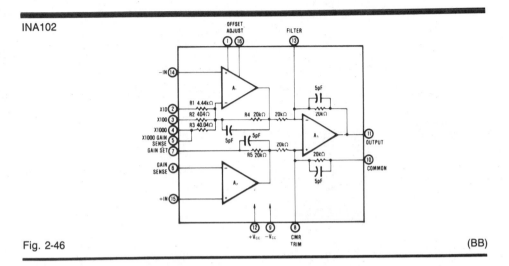

Fig. 2-46 (BB)

and the matching must be maintained over temperature and time in order to retain good common-mode rejection.

All of the internal resistors are made of thin-film nichrome on the integrated circuit. The critical resistors are laser-trimmed to provide the desired high gain accuracy and common-mode rejection. Nichrome ensures long-term stability and provides excellent TCR and TCR tracking. This provides gain accuracy and common-mode rejection when the INA102 is operated over wide temperature ranges.

Using the INA102

Figure 2-46 shows the simplest configuration of the INA102. The output voltage is a function of the differential input voltage times the gain.

A gain of 1, 10, 100, or 1000 is selected by programming pins 2 through 7 (see Table 2-I). Notice that for the gain of 1000, a special gain sense is provided to preserve accuracy. Although this is not always required, gain errors caused by external resistance in series with the low value 40.04 Ω internal gain set resistor are thus eliminated.

Other gains between 1 and 10, 10 and 100, and 100 and 1000 can also be obtained by connecting an external resistor between pin 6 and either pin 2, 3, or 4, respectively (see Fig. 2-54 for application).

$G = 1 + (40/R_G)$ where R_G is the total resistance between the two inverting inputs of the input op amps. At high gains, where the value of R_G becomes small, additional resistance (i.e., relays or sockets) in the R_G circuit will contribute to a gain error. Care should be taken to minimize this effect.

Optional Offset Adjustment Procedure

It is sometimes desirable to null the input and/or output offset to achieve higher accuracy. The quality of the potentiometer will affect the results; therefore, choose one with good temperature and mechanical-resistance stability.

INA102

SPECIFICATIONS ELECTRICAL

At $T_A = +25°C$ with ±15VDC power supply and in circuit of Figure 2 unless otherwise noted.

MODEL	CONDITIONS	INA102AG MIN	TYP	MAX	INA102CG MIN	TYP	MAX	UNITS
GAIN								
Range of Gain		1		1000	*		*	V/V
Gain Equation, External, ±20%			$G = 1 + (40k/R_G)^{(1)}$			*		V/V
Error, DC: G = 1	$T_A = +25°C$			0.1			0.05	%
G = 10	$T_A = +25°C$			0.1			0.05	%
G = 100	$T_A = +25°C$			0.25			0.15	%
G = 1000	$T_A = +25°C$			0.75			0.5	%
G = 1	$T_A = T_{MIN}$ to T_{MAX}			0.16			0.08	%
G = 10	$T_A = T_{MIN}$ to T_{MAX}			0.19			0.11	%
G = 100	$T_A = T_{MIN}$ to T_{MAX}			0.37			0.21	%
G = 1000	$T_A = T_{MIN}$ to T_{MAX}			0.93			0.62	%
Gain Temp. Coefficient								
G = 1				10			5	ppm/°C
G = 10				15			10	ppm/°C
G = 100				20			15	ppm/°C
G = 1000				30			20	ppm/°C
Nonlinearity, DC								
G = 1	$T_A = +25°C$			0.03			0.01	% of FS
G = 10	$T_A = +25°C$			0.03			0.01	% of FS
G = 100	$T_A = +25°C$			0.05			0.02	% of FS
G = 1000	$T_A = +25°C$			0.1			0.05	% of FS
G = 1	$T_A = T_{MIN}$ to T_{MAX}		*	0.045			0.015	% of FS
G = 10	$T_A = T_{MIN}$ to T_{MAX}			0.045			0.015	% of FS
G = 100	$T_A = T_{MIN}$ to T_{MAX}			0.075			0.03	% of FS
G = 1000	$T_A = T_{MIN}$ to T_{MAX}			0.15			0.1	% of FS
RATED OUTPUT								
Voltage	$R_L = 10k\Omega$	±($|V_{CC}|$−2.5)			*	*		V
Current		±1			*	*		mA
Short-Circuit Current			2			*		mA
Output Impedance								
G = 1000			0.1			*		Ω
INPUT								
OFFSET VOLTAGE								
Initial Offset[2]	$T_A = +25°C$			±300 ±300/G			±100 ±200/G	µV
vs Temperature				±5 ±10/G			±2 ±5/G	µV/°C
vs Supply				±40 ±50/G			±10 ±20/G	µV/V
vs Time			±(20 + 30/G)			*		µV/mo
BIAS CURRENT								
Initial Bias Current (each input)	$T_A = T_{MIN}$ to T_{MAX}		±25	50		6	30	nA
vs Temperature			±0.1			*		nA/°C
vs Supply			±0.1			*		nA/V
Initial Offset Current	$T_A = T_{MIN}$ to T_{MAX}		±2.5	±15		±2.5	±10	nA
vs Temperature			±0.1			*		nA/°C
IMPEDANCE								
Differential			$10^{10}\|2$			*		Ω ‖ pF
Common-mode			$10^{10}\|2$			*		Ω ‖ pF
VOLTAGE RANGE								
Range, Linear Response	$T_A = T_{MIN}$ to T_{MAX}	±($|V_{CC}|$−2.5)			*			V
CMR with 1kΩ Source Imbalance								
G = 1	DC to 60 Hz	80	94		90	94		dB
G = 10	DC to 60 Hz	80	100		90	100		dB
G = 10 to 1000	DC to 60 Hz	80	100		90	100		dB
NOISE								
Input Voltage Noise								
$f_B = 0.01Hz$ to 10Hz			0.1			*		µV, p-p
Density, G = 1000								
$f_O = 10Hz$			30			*		nV/√Hz
$f_O = 100Hz$			25			*		nV/√Hz
$f_O = 1kHz$			25			*		nV/√Hz
Input Current Noise								
$f_B = 0.01Hz$ to 10Hz			25			*		pA p-p
Density: $f_O = 10Hz$			0.3			*		pA/√Hz
$f_O = 100Hz$			0.2			*		pA/√Hz
$f_O = 1kHz$			0.15			*		pA/√Hz
DYNAMIC RESPONSE								
Small Signal ±3dB Flatness	$V_{OUT} = 0.1V_{RMS}$							
G = 1			300			*		kHz
G = 10			30			*		kHz
G = 100			3			*		kHz
G = 1000			0.3			*		kHz

Fig. 2-47

(BB)

INA102
ELECTRICAL (CONT)

MODEL		INA102AG			INA102CG			UNITS
	CONDITIONS	MIN	TYP	MAX	MIN	TYP	MAX	
Small Signal, ±1% Flatness	$V_{OUT} = 0.1V_{RMS}$							
G = 1			30			•		kHz
G = 10			3			•		kHz
G = 100			0.3			•		kHz
G = 1000			0.03			•		kHz
Full Power, G = 1 to 100	$V_{OUT}=10V, R_L=10k\Omega$	2.4	3		•			kHz
Slew Rate, G = 1 to 100	$V_{OUT}=10V, R_L=10k\Omega$	0.15	0.2		•			V/µsec
Settling Time	$R_L=10k\Omega, C_L=100pF$							
0.1%: G = 1	10V step		50			•		µsec
G = 100			360					µsec
G = 1000			3300			•		µsec
0.01%: G = 1	10V step		60					µsec
G = 100			500			•		µsec
G = 1000			4500			•		µsec
POWER SUPPLY								
Rated Voltage			±15			•		V
Voltage Range		±3.5		±18	•		•	V
Quiescent Current[3]	$V_O = 0V$, $T_A = T_{MIN}$ to T_{MAX}		±500	±750		•	•	µA
TEMPERATURE RANGE								
Specification		−25		+85	•		•	°C
Operation	$R_L > 50k\Omega$[3]	−25		+85	•		•	°C
Storage		−65		+150	•		•	°C

*Specifications same as for INA102AG.

NOTES: (1) The internal gain set resistors have an absolute tolerance of ±20%; however, their tracking is 50ppm/°C. R_G will add to the gain error if gains other than 1, 10, 100 or 1000 are set externally. (2) Adjustable to zero at any one time. (3) At high temperature, output drive current is limited. An external buffer can be used if required.

The information provided herein is believed to be reliable; however, BURR-BROWN assumes no responsibility for inaccuracies or omissions. BURR-BROWN assumes no responsibility for the use of this information, and all use of such information shall be entirely at the user's own risk. Prices and specifications are subject to change without notice. No patent rights or licenses to any of the circuits described herein are implied or granted to any third party. BURR-BROWN does not authorize or warrant any BURR-BROWN product for use in life support devices and/or systems.

PIN CONFIGURATION

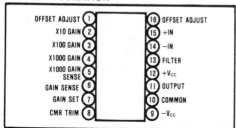

PRICES

Quantity	AG	CG
1-24	$13.95	$18.40
25-99	11.15	15.85
100's	7.95	11.35

MECHANICAL

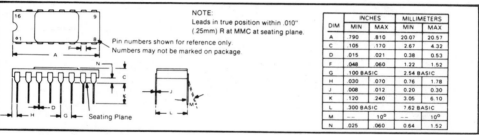

ABSOLUTE MAXIMUM RATINGS

Supply .... ±18V
Input Voltage Range .... ±V_{CC}
Operating Temperature
    Range .... −25°C to +85°C
Storage Temperature
    Range .... −65°C to +150°C
Lead Temperature
    (soldering 10 seconds).......... +300°C
Output Short-Circuit
    Duration .... Continuous to ground

ORDERING INFORMATION

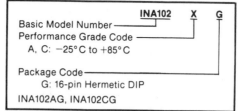

Basic Model Number
Performance Grade Code
    A, C: −25°C to +85°C

Package Code
    G: 16-pin Hermetic DIP
INA102AG, INA102CG

Fig. 2-48

(BB)

INA102

At +25°C and in circuit of Figure 2 unless otherwise noted.

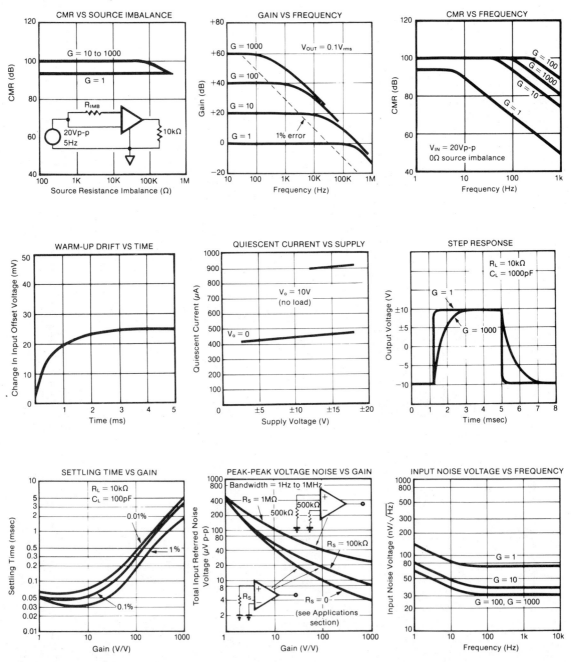

Fig. 2-49

(BB)

INA102

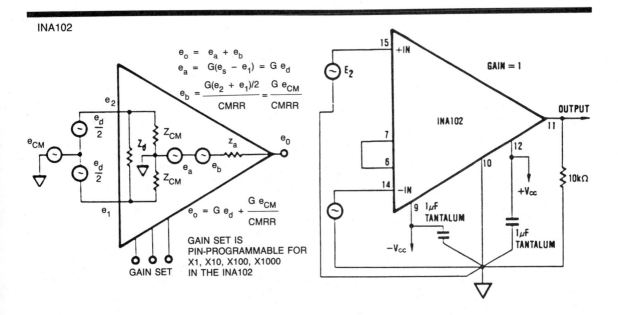

Fig. 2-50A & B   Model of an instrumentation amplifier.   (BB)

The optional offset null capabilities are shown in Fig. 2-51. R_4 adjustment affects only the input stage component of the offset voltage. Note that the null condition will be disturbed when the gain is changed. Also, the input drift will be affected by approximately 0.31 $\mu V/^\circ$ C per 100 μV of input offset voltage that is trimmed. Therefore, care should be taken when considering use of the control for removal of other sources of offset. Output offset correction can be accomplished with A_1, R_1, R_2, and R_3, by applying a voltage to Common (pin 10) through a buffer amplifier. This buffer limits the resistance in series with pin 10 to minimize CMR error. Resistance above 0.1 Ω will cause the common-mode rejection to fall below 100 dB. Be certain to keep this resistance low.

It is important to not exceed the input amplifier's dynamic range. The amplified differential input signal and its associated common-mode voltage should not cause the output of A_1 or A_2 to exceed approximately ± 12 V with ± 15 V supplies or nonlinear operation will result. To protect against moisture, especially in high gain, sealing compound may be used. Current injected into the offset pins should be minimized.

GAIN	CONNECT PINS
1	6 to 7
10	2 to 6 and 7
100	3 to 6 and 7
1000	4 to 7 and separately 5 to 6

Table 2-1   Pin-programmable gain connections.

INA102

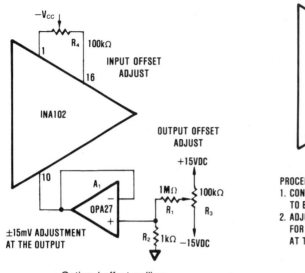

Optional offset nulling

INA102

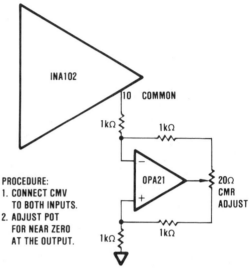

PROCEDURE:
1. CONNECT CMV TO BOTH INPUTS.
2. ADJUST POT FOR NEAR ZERO AT THE OUTPUT.

Optional circuit for externally trimming CMR.

Fig. 2-51   (BB)

Fig. 2-52   (BB)

INA102

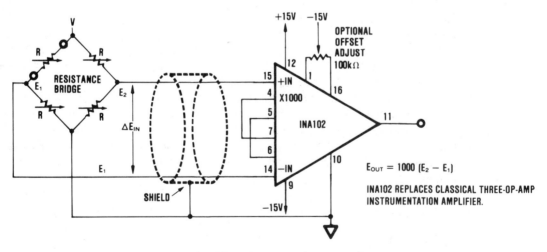

$E_{OUT} = 1000 (E_2 - E_1)$

INA102 REPLACES CLASSICAL THREE-OP-AMP INSTRUMENTATION AMPLIFIER.

Amplification of a differential voltage from a resistance bridge.

Fig. 2-53   (BB)

INA102

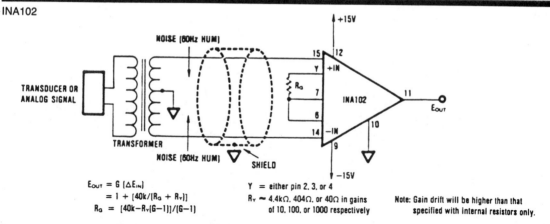

$$E_{OUT} = G \, [\Delta E_{IN}]$$
$$= 1 + [40k/(R_G + R_Y)]$$
$$R_G = [40k - R_Y(G-1)]/(G-1)$$

Y = either pin 2, 3, or 4
$R_Y \approx 4.4k\Omega$, 404Ω, or 40Ω in gains of 10, 100, or 1000 respectively

Note: Gain drift will be higher than that specified with internal resistors only.

Fig. 2-54   Amplification of a transformer-coupled analog signal using external gain set.   (BB)

INA102

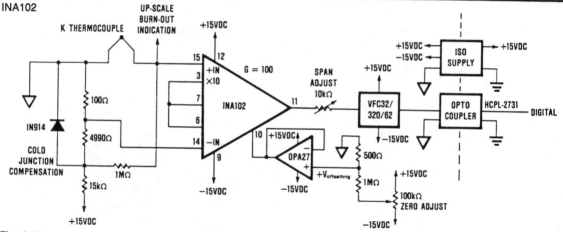

Fig. 2-55   Isolated thermocouple amplifier with cold junction compensation.   (BB)

INA102

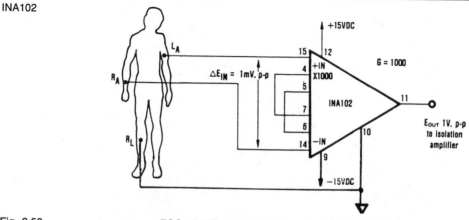

Fig. 2-56   ECG amplifier or recorder preamp for biological signals.   (BB)

INA102

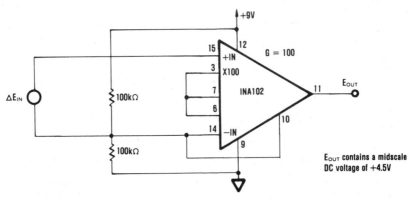

Fig. 2-57 Single supply low power instrumentation amplifier. (BB)

INA102

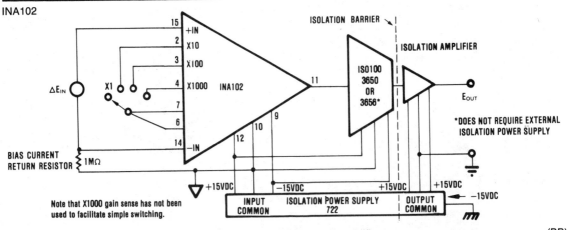

Fig. 2-58 Precision isolated instrumentation amplifier. (BB)

INA102

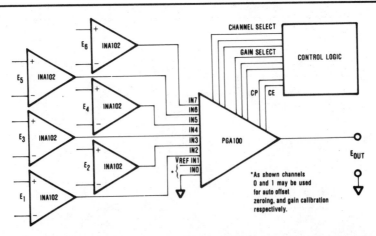

Fig. 2-59 Multiple channel precision instrumentation amplifier with programmable gain. (BB)

INA102

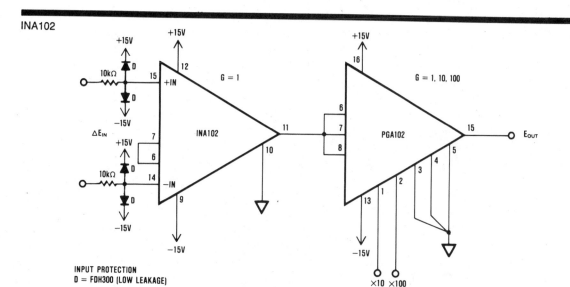

Fig. 2-60   Programmable-gain instrumentation amplifier using the INA102 and PGA102.   (BB)

INA102

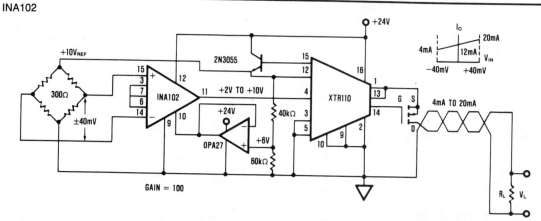

Fig. 2-61   4 mA to 20 mA bridge transmitter using single supply instrumentation amplifier.   (BB)

INA102

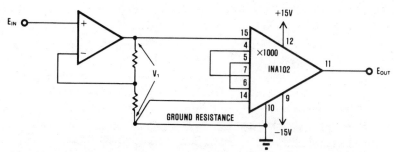

Fig. 2-62   Ground resistance loop eliminator (INA102 senses and amplifies V_1 accurately).   (BB)

INA102

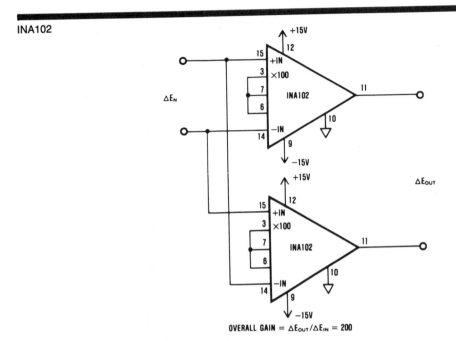

OVERALL GAIN = $\Delta E_{OUT}/\Delta E_{IN}$ = 200

Fig. 2-63   Differential input/differential output amplifier (twice the gain of one INA).   (BB)

INA102

CONTROL	S1	S2	S3	S4	S5	MODE
0	Closed	Closed	Open	Open	Closed	Auto-Zeroing
1	Open	Open	Closed	Closed	Open	Signal Amplification

Fig. 2-64   Auto-zeroing instrumentation amplifier circuit.   (BB)

Optional Filtering

The INA102 has provisions for accomplishing filtering with one external capacitor between pins 11 and 13. This single-pole filter can be used to reduce noise outside the signal bandwidth, but with degradation to ac CMR.

When it is important to preserve CMR versus frequency (especially at 60 Hz), two capacitors should be used. The additional capacitor is connected between pins 8 and 10. This will maintain a balance of impedances in the output stage. Either of these capacitors could also be trimmed slightly to maximize CMR, if desired. Note that their ratio tracking will affect CMR over temperature.

Optional Common-Mode Rejection Trim

The INA102 is laser-adjusted during manufacturing to assure high CMR. However, if desired, a small resistance can be added in series with pin 10 to trim the CMR to an improved level. Depending upon the nature of the internal imbalances, either a positive or negative resistance value could be required. The circuit shown in Fig. 2-52 acts as a bipolar potentiometer and allows easy adjustment of CMR.

Typical Applications

Many applications of instrumentation amplifiers involve the amplification of low level differential signals from bridges and transducers such as strain gauges, thermocouples, and RTD's. Some of the important parameters include common-mode rejection (differential cancellation of common-mode offset and noise, see Fig. 2-50A), input impedance, offset voltage and drift, gain accuracy, linearity, and noise. The INA102 accomplishes all of these with high precision at surprisingly low quiescent current. However, in higher gains (>10) with high source impedances (>100 kΩ), the bias current can cause a large offset at the output. This can saturate the output unless the source impedance is separated, e.g., two 500 kΩ paths instead of one 1 MΩ unbalanced input. The input offset current times 500 kΩ will then generate a small dc voltage error.

Figures 2-53 through 2-59 show some typical applications circuits.

LOUDSPEAKER AMPLIFIER

The TEA7031 is a 28-pin DIL integrated circuit especially designed to be used as loudspeaker amplifier.
Functions implemented on the chip include:

- ☐ Loudspeaker amplifier,
- ☐ Anti-acoustic feed-back system (anti-larsen),
- ☐ Direct microcomputer supply,
- ☐ Switching regulator.

These functions are electrically separated and may be used individually. One of the main applications is the telephone set with loudspeaker. In this configuration, the circuit is used in conjunction with TEA7030.

Loudspeaker Amplifier

- ☐ Dc voltage : from 2.5 V to 7 V
- ☐ Supply current : less than 1.5 mA without output current
- ☐ High output voltage swing : 3 Vpp on a 50 Ω loudspeaker with only 3 Vdc supply.
- ☐ The gain of the amplifier is programmable linearly or in steps of 6 dB
- ☐ When the required output energy becomes higher than the energy available by the power supply, an automatic gain control system will reduce the gain to avoid distortion.

High Efficiency Anti-Acoustic Feed-Back System

- ☐ Adjustable as a function of the mechanical feed-back
- ☐ An original system will distinguish between voice and other signals thus preventing the amplifier to switch off in the presence of background room noise signals.

Microcomputer Supply

- ☐ Dc voltage : from 2 V to 5 V
- ☐ Reset and halt signals available
- ☐ A Ring Detection Signal (RDS) is available to allow the circuit to be used as a ringer for a telephone set.

Switching Regulator

- ☐ Low operating supply voltage
- ☐ If the circuit is supplied by a high voltage energy source, e.g., 24, (which is normally the case while ringing signal is being received), due to high impedance characteristic of the circuit, the available current will be insufficient for satisfactory circuit operation. Under this condition, the on-chip SWITCHING REGULATOR will convert the available high voltage into a low voltage (e.g., 3 V) and will provide the required amount of current for high efficiency circuit operation.

Switching Supply

☐ This supply allows the circuit to be used for a ringer.

☐ It converts the high voltage to high current on the loudspeaker and powers the IC. So the same circuit can be used both as voice amplifier and ringer amplifier.

☐ When the switching regulator is operating properly and the microcomputer power supply is satisfactory, the circuit will send an active "RDS" signal to the microcomputer to instruct it to generate a melody signal. Conversely, the microcomputer will return the melody signal which is then processed internally by the TEA7031 and applied to the loudspeaker.

TEA7031

PIN ASSIGNMENT

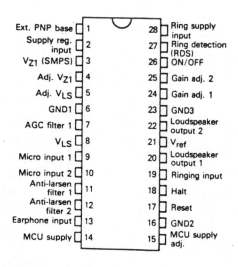

Pin		Pin	
Ext. PNP base	1	28	Ring supply input
Supply reg. input	2	27	Ring detection (RDS)
V_{Z1} (SMPS)	3	26	ON/OFF
Adj. V_{Z1}	4	25	Gain adj. 2
Adj. V_{LS}	5	24	Gain adj. 1
GND1	6	23	GND3
AGC filter 1	7	22	Loudspeaker output 2
V_{LS}	8	21	V_{ref}
Micro input 1	9	20	Loudspeaker output 1
Micro input 2	10	19	Ringing input
Anti-larsen filter 1	11	18	Halt
Anti-larsen filter 2	12	17	Reset
Earphone input	13	16	GND2
MCU supply	14	15	MCU supply adj.

Fig. 2-65

(TH)

TEA7031

APPLICATION FOR TELEPHONE SET

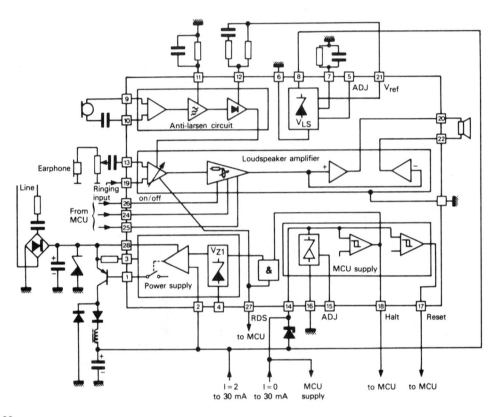

Fig. 2-66

(TH)

CRT DEFLECTION AMPLIFIER

In order to obtain more ideal complex load performance, a high-current wideband width IC Op Amp. (L165), is interfaced with a DMOS complimentary output stage. The L165's current capability, is enhanced by DMOS's excellent source to load isolation ability. A 0.5 V peak to peak input signal develops up to a ± 10 volts 8 amps output swing. The feedback is sensed by the 0.06 ohm grounded resistor in series with the deflection coil.

VN1210N1, VP1210N1

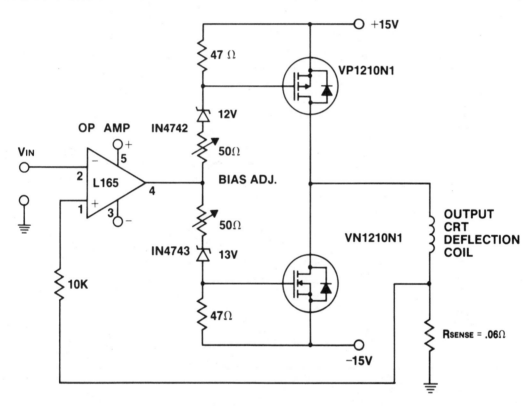

Fig. 2-67

(SU)

125 WATT RMS AUDIO AMPLIFIER

The keynotes for this circuit are simplicity and small size. The dual Op Amp provides 26 dB voltage gain to inverting amplifiers with ± 19 V output swing at pins 1 & 7, with 100 kHz power bandwidth. The complimentary configured P & N DMOS output stage produces over 7 amperes into a 4 ohm load. Maximum transistor dissipation is 40 watts per device allowing use of TO-3 packages. This amplifier should provide the "tube type" overdrive characteristics with low THD. In addition, this design easily powers ac or dc and transformer coupled circuits.

VN1106N1, VP1206N1

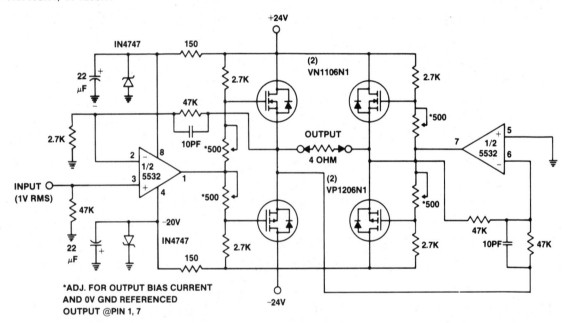

Fig. 2-68

(SU)

3
Oscillator, Timer, Counter, Clock, and Multiplier/Divider Circuits

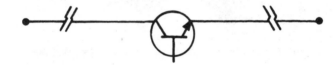

PRECISION WAVEFORM GENERATOR/VCO

The ICL8038 Waveform Generator is a monolithic integrated circuit capable of producing high accuracy sine, square, triangular, sawtooth and pulse waveforms with a minimum of external components. The frequency (or repetition rate) can be selected externally from .001 Hz to more than 300 kHz using either resistors or capacitors, and frequency modulation and sweeping can be accomplished with an external voltage. The ICL8038 is fabricated with advanced monolithic technology, using Schottky-barrier diodes and thin film resistors, and the output is stable over a wide range of temperature and supply variations. These devices may be interfaced with phase locked loop circuitry to reduce temperature drift to less than 250 ppm/° C.

Features

☐ Low Frequency Drift With Temperature—250 ppm/° C
☐ Simultaneous Sine, Square, and Triangle Wave Outputs
☐ Low Distortion—1% (Sine Wave Output)
☐ High Linearity—0.1% (Triangle Wave Output)
☐ Wide Operating Frequency Range—0.001 Hz to 300 kHz
☐ Variable Duty Cycle—2% to 98%
☐ High Level Outputs—TTL to 28 V
☐ Easy to Use—just a handful of external components required

ICL8038

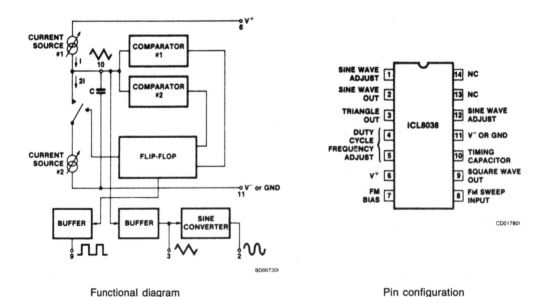

Functional diagram

Pin configuration

Fig. 3-1

(IN)

Definition of Terms

☐ Supply Voltage (V_{SUPPLY}. The total supply voltage from V+ to V−
☐ Supply Current. The supply current required from the power supply to operate the device, excluding load currents and the currents through R_A and R_B.
☐ Frequency Range. The frequency range at the square wave output through which circuit operation is guaranteed.
☐ Sweep FM Range. The ratio of maximum frequency to minimum frequency which can be obtained by applying a sweep voltage to pin 8. For correct operation, the sweep voltage should be within the range

$$(\tfrac{2}{3}\ V_{SUPPLY} + 2\ V) < V_{SWEEP} < V_{SUPPLY}$$

ICL8038

ELECTRICAL CHARACTERISTICS ($V_{SUPPLY} = \pm 10V$ or $+20V$, $T_A = 25°C$, $R_L = 10k\Omega$, Test Circuit Unless Otherwise Specified)

SYMBOL	GENERAL CHARACTERISTICS	8038CC			8038BC(BM)			8038AC(AM)			UNIT
		MIN	TYP	MAX	MIN	TYP	MAX	MIN	TYP	MAX	
V_{SUPPLY}	Supply Voltage Operating Range										
V+	Single Supply	+10		+30	+10		30	+10		30	V
V+, V−	Dual Supplies	±5		±15	±5		±15	±5		±15	V
I_{SUPPLY}	Supply Current ($V_{SUPPLY} = \pm 10V$)[2]										
	8038AM, 8038BM					12	15		12	15	mA
	8038AC, 8038BC, 8038CC		12	20		12	20		12	20	mA
FREQUENCY CHARACTERISTICS (all waveforms)											
f_{max}	Maximum Frequency of Oscillation	100			100			100			kHz
f_{sweep}	Sweep Frequency of FM Input		10			10			10		kHz
	Sweep FM Range[3]		35:1			35:1			35:1		
	FM Linearity 10:1 Ratio		0.5			0.2			0.2		%
$\Delta f/\Delta T$	Frequency Drift With Temperature[5] 8038 AC, BC, CC 0°C to 70°C		.250			180			110		ppm/°C
	8038 AM, BM, −55°C to 125°C						350			250	
$\Delta f/\Delta V$	Frequency Drift With Supply Voltage (Over Supply Voltage Range)		0.05			0.05			0.05		%/V
OUTPUT CHARACTERISTICS											
I_{OLK}	**Square-Wave** Leakage Current ($V_9 = 30V$)			1			1			1	μA
V_{SAT}	Saturation Voltage ($I_{SINK} = 2mA$)		0.2	0.5		0.2	0.4		0.2	0.4	V
t_r	Rise Time ($R_L = 4.7k\Omega$)		180			180			180		ns
t_f	Fall Time ($R_L = 4.7k\Omega$)		40			40			40		ns
ΔD	Typical Duty Cycle Adjust (Note 6)	2		98	2		98	2		98	%
$V_{TRIANGLE}$	**Triangle/Sawtooth/Ramp** Amplitude ($R_{TRI} = 100k\Omega$)	0.30	0.33		0.30	0.33		0.30	0.33		$\times V_{SUPPLY}$
	Linearity		0.1			0.05			0.05		%
Z_{OUT}	Output Impedance ($I_{OUT} = 5mA$)		200			200			200		Ω
V_{SINE}	**Sine-Wave** Amplitude ($R_{SINE} = 100k\Omega$)	0.2	0.22		0.2	0.22		0.2	0.22		$\times V_{SUPPLY}$
THD	THD ($R_S = 1M\Omega$)[4]		2.0	5		1.5	3		1.0	1.5	%
THD	THD Adjusted (Use Figure 6)		1.5			1.0			0.8		%

NOTES: 2. R_A and R_B currents not included.
3. $V_{SUPPLY} = 20V$; R_A and $R_B = 10k\Omega$, $f \cong 10kHz$ nominal; can be extended 1000 to 1. See Figures 7a and 7b.
4. 82kΩ connected between pins 11 and 12, Triangle Duty Cycle set at 50%. (Use R_A and R_B.)
5. Figure 3, pins 7 and 8 connected, $V_{SUPPLY} = \pm 10V$. See Typical Curves for T.C. vs V_{SUPPLY}.

Fig. 3-2

(IN)

ICL8038

Test Conditions

PARAMETER		R_A	R_B	R_L	C_1	SW_1	MEASURE
Supply Current		10kΩ	10kΩ	10kΩ	3.3nF	Closed	Current into Pin 6
Sweep FM Range[1]		10kΩ	10kΩ	10kΩ	3.3nF	Open	Frequency at Pin 9
Frequency Drift with Temperature		10kΩ	10kΩ	10kΩ	3.3nF	Closed	Frequency at Pin 3
Frequency Drift with Supply Voltage[2]		10kΩ	10kΩ	10kΩ	3.3nF	Closed	Frequency at Pin 9
Output Amplitude:	Sine	10kΩ	10kΩ	10kΩ	3.3nF	Closed	Pk-Pk output at Pin 2
(Note 4)	Triangle	10kΩ	10kΩ	10kΩ	3.3nF	Closed	Pk-Pk output at Pin 3
Leakage Current (off)[3]		10kΩ	10kΩ		3.3nF	Closed	Current into Pin 9
Saturation Voltage (on)[3]		10kΩ	10kΩ		3.3nF	Closed	Output (low) at Pin 9
Rise and Fall Times (Note 5)		10kΩ	10kΩ	4.7kΩ	3.3nF	Closed	Waveform at Pin 9
Duty Cycle Adjust:	MAX	50kΩ	~1.6kΩ	10kΩ	3.3nF	Closed	Waveform at Pin 9
(Note 5)	MIN	~25kΩ	50kΩ	10kΩ	3.3nF	Closed	Waveform at Pin 9
Triangle Waveform Linearity		10kΩ	10kΩ	10kΩ	3.3nF	Closed	Waveform at Pin 3
Total Harmonic Distortion		10kΩ	10kΩ	10kΩ	3.3nF	Closed	Waveform at Pin 2

NOTES: 1. The hi and lo frequencies can be obtained by connecting pin 8 to pin 7 (f_{hi}) and then connecting pin 8 to pin 6 (f_{lo}). Otherwise apply Sweep Voltage at pin 8 (2/3 V_{SUPPLY} + 2V) ≤ V_{SWEEP} ≤ V_{SUPPLY} where V_{SUPPLY} is the total supply voltage. In Figure 7b, pin 8 should vary between 5.3V and 10V with respect to ground.
2. 10V ≤ V^+ ≤ 30V, or ±5V ≤ V_{SUPPLY} ≤ ±15V.
3. Oscillation can be halted by forcing pin 10 to +5 volts or −5 volts.
4. Output Amplitude is tested under static conditions by forcing pin 10 to 5.0V then to −5.0V.
5. Not tested; for design purposes only.

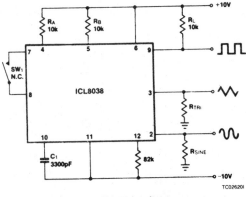

Fig. 3-3   Test circuit   (IN)

FM Linearity. The percentage deviation from the best-fit straight line on the control voltage versus output frequency curve.

Output Amplitude. The peak-to-peak signal amplitude appearing at the outputs.

Saturation Voltage. The output voltage at the collector of Q23 when this transistor is turned on. It is measured for a sink current of 2 mA.

Rise and Fall Times. The time required for the square wave output to change from 10% to 90%, or 90% to 10%, of its final value.

Triangle Waveform Linearity. The percentage deviation from the best-fit straight line on the rising and falling triangle waveform.

Total Harmonic Distortion. The total harmonic distortion at the sine-wave output.

ICL8038

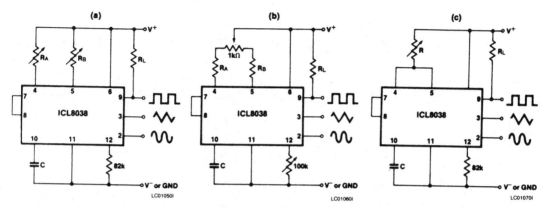

Possible connections for the external timing resistors

Fig. 3-4

(IN)

ICL8038

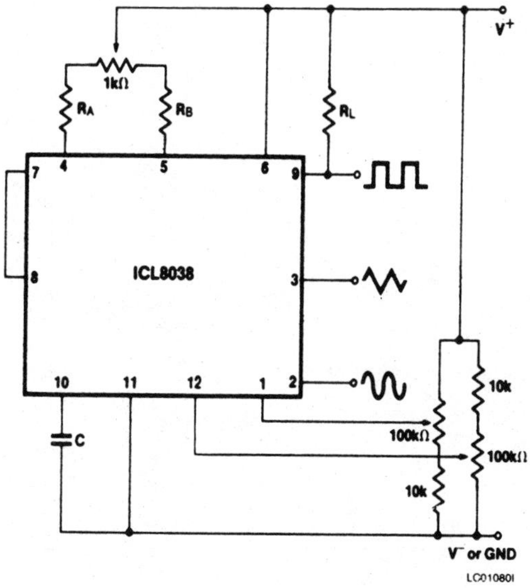

Connection to achieve minimum sine-wave distortion

Fig. 3-5

(IN)

ICL8038

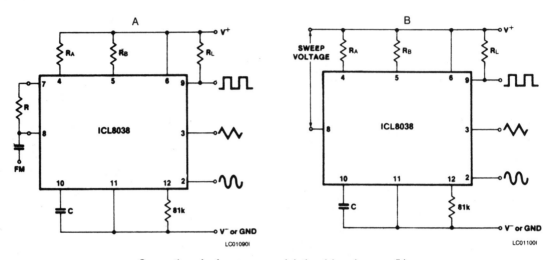

Connections for frequency modulation (a) and sweep (b)

Fig. 3-6 (IN)

ICL8038

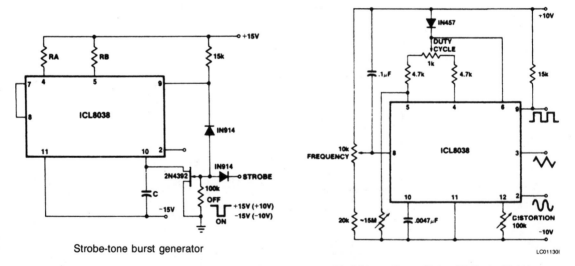

Strobe-tone burst generator

Variable audio oscillator, 20 Hz to 20 kHz

Fig. 3-7 (IN)

ICL8038

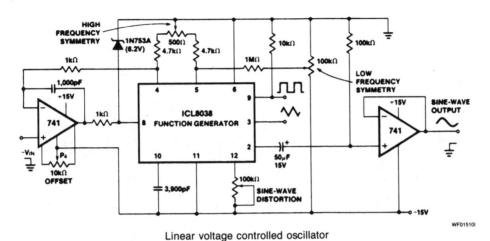

Linear voltage controlled oscillator

WF01510I

Fig. 3-7. Continued from page 122.   (IN)

ICL8038

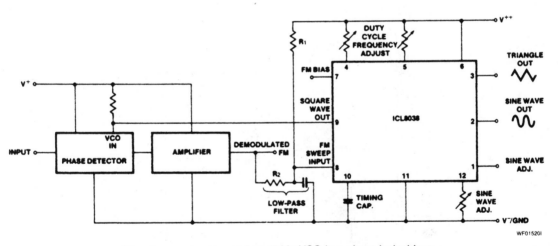

WF01520I

Waveform generator used as stable VCO in a phase-locked loop

Fig. 3-8   (IN)

TL071, TL072, TAA761C

(0.5 Hz) SQUARE WAVE OSCILLATOR

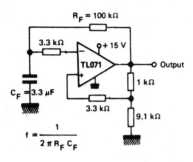

$$f = \frac{1}{2 \pi R_F C_F}$$

PULSE GENERATOR

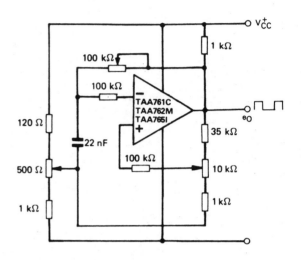

QUADRATURE OSCILLATOR

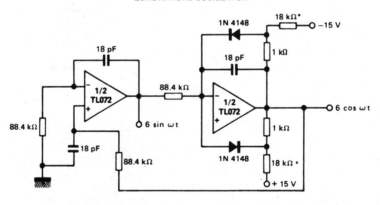

* These resistor values may be adjusted for a symmetrical output.

Fig. 3-9

(TH)

LM111, LM139

CRYSTAL OSCILLATOR

CRYSTAL CONTROLLED OSCILLATOR

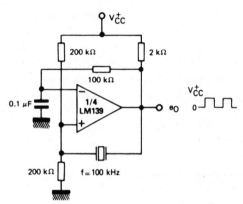

TWO-DECADE HIGH-FREQUENCY VCO

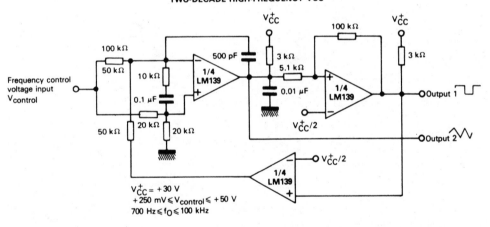

$V_{CC}^+ = +30$ V
$+250$ mV $\leqslant V_{control} \leqslant +50$ V
700 Hz $\leqslant f_O \leqslant 100$ kHz

Fig. 3-10

(TH)

ULTRA-SIMPLE SATURABLE CORE OSCILLATOR

Low to medium, minimum-component oscillators and frequency seen in multi-voltage portable instruments, and electroluminescent/backlit LCD displays. The use of the VP & VN01 transistors eliminated the usual base drive windings on T1, and associated support components needed in bipolar circuits.

When turned on, the circuit selects a state from positive feedback and latches until the inductor saturates. As the inductor saturates the current in the 10 ohm resistor increases rapidly until the amplifier supply voltage drops below the FET threshold. As the FETs start to turn off, the collapsing field in the inductor flips this circuit quickly to the other stable state.

The two transistor circuit with only 4 components, is simpler for higher power applications.

VN0104N3, VC0106N6

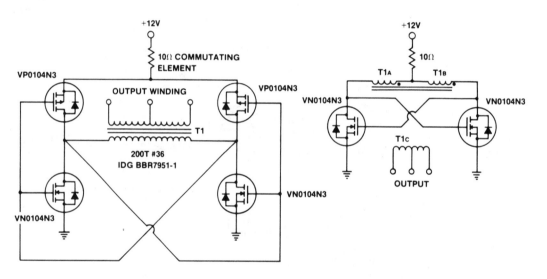

Fig. 3-11

(SU)

PRECISION QUADRATURE OSCILLATOR

The 4423 does not require any external component to obtain a 20 kHz quadrature oscillator. The connection diagram is as shown in Fig. 3-12.

For resistor programmable frequencies in the 2 kHz to 20 kHz frequency range, the connection diagram is shown in Fig. 3-13. Note that only two resistors of equal value are required. The resistor R can be expressed by,

$$R = \frac{3.785\ f}{42.05 - 2\ f} \qquad \text{where, R in k}\Omega \\ \qquad\qquad\qquad f \text{ in kHz}$$

For oscillator frequencies below 2000 Hz, use of two capacitors of equal value and two resistors of equal value as shown in Fig. 3-14 is recommended. Connections shown in Fig. 3-14 can be used to get oscillator frequency in the 0.002 Hz to 20 kHz range.

The frequency f can be expressed by:

$$f = \frac{42.05\ R}{(C + 0.001)\ (3.785 + 2\ R)}$$

where, f is in Hz
　　　C is in μF
and R is in kΩ.

For best results, the capacitor values shown in Table 3-1 should be selected with respect to their frequency ranges.

After selecting the capacitor for a particular frequency the value of the required resistor can be obtained by using the resistor selection curve shown in Fig. 3-15 or by the expression:

$$R = \frac{3.785\ f\ (C + 0.001)}{42.05 - 2\ f\ (C + 0.001)}$$

where, R is in kΩ
　　　f is in Hz
and C is in μF.

4423

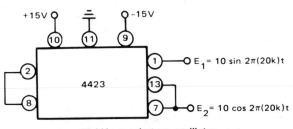

20 kHz quadrature oscillator.

Fig. 3-12   (BB)

4423

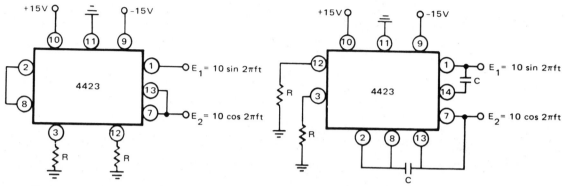

Resistor programmable quadrature oscillator.

Quadrature oscillator programmable to 0.002 Hz.

Fig. 3-13   (BB)   Fig. 3-14   (BB)

4423

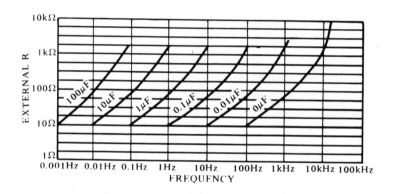

Resistor selection curves.

Fig. 3-15   (BB)

Capacitor values according to frequency range.

f	20 kHz to 2 kHz	2 kHz to 200 Hz	200 Hz to 20 Hz
C	0	$0.01\mu F$	$0.1\mu F$
20 Hz to 2 Hz	2 Hz to 0.2 Hz	0.2 Hz to 0.02 Hz	0.02 Hz to 0.002 Hz
$1\mu F$	$10\mu F$	$100\mu F$	$1000\mu F$

Table 3-1

LM111

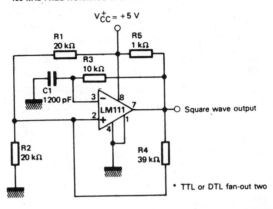

100 kHz FREE RUNNING MULTIVIBRATOR

Fig. 3-16   (TH)

LM101A, LM201A, LM301A

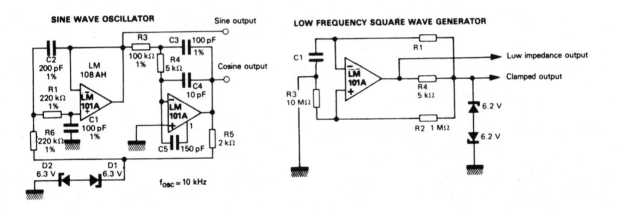

SINE WAVE OSCILLATOR

f_{osc} = 10 kHz

LOW FREQUENCY SQUARE WAVE GENERATOR

Fig. 3-17

555 TIMER CIRCUIT

The NE555/SE555 monolithic timing circuit is a highly stable controller capable of producing accurate time delays or oscillations. Additional terminals are provided for triggering or resetting if desired. In the time delay mode of operation, the time is precisely controlled by one external resistor and capacitor.

For astable operation as an oscillator, the free running frequency and the duty cycle are both accurately controlled with two external resistors and one capacitor.

The circuit may be triggered and reset on falling waveforms, and the output structure can source or sink up to 200 mA or drive TTL circuits.

☐ Timing from microseconds through hours.
☐ Operates in both astable and monostable modes.
☐ Adjustable duty cycle.
☐ High current output can source or sink 200 mA.
☐ Temperature stability of 0.005% per °C.

NE555, SE555

PIN ASSIGNMENTS
(Top views)

CB-11

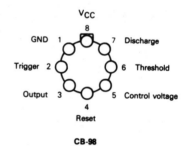

ORDERING INFORMATION

Hi-Rel versions available - See chapter 14

PART NUMBER	TEMPERATURE RANGE	PACKAGE			
		DP	DG	FP	H
NE555	0°C to + 70°C	•	•	•	•
SE555	−55°C to + 125°C	•	•		•
NE555I	−40°C to + 85°C	•	•		

Examples : NE555DP, NE555IDG

CB-98
CB-342

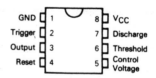

Fig. 3-18

(TH)

NE555, SE555

MAXIMUM RATINGS

Rating		Symbol	Value	Unit
Power supply voltage		V_{CC}	18	V
Output current		I_O	200	mA
Power dissipation		P_{tot}	600	mW
Operating free-air temperature range		T_{oper}		°C
	SE555		−55 to +125	
	NE555		0 to +70	
	NE555I		−40 to +85	
Storage temperature range		T_{stg}	−65 to +150	°C

SCHEMATIC DIAGRAM

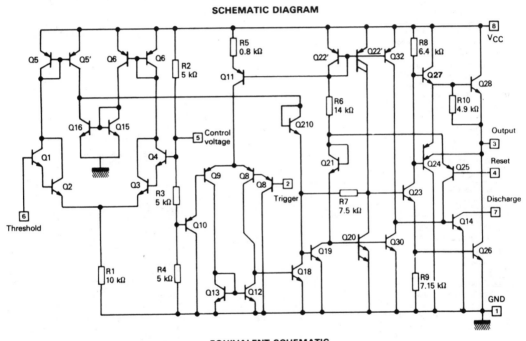

EQUIVALENT SCHEMATIC

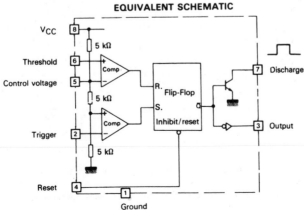

Fig. 3-18. Continued through page 136.

(TH)

NE555, SE555

ELECTRICAL CHARACTERISTICS

$T_{amb} = +25°C$, $V_{CC} = +5$ V to $+15$ V
(Unless otherwise specified)

Characteristic	Symbol	SE555			NE555, NE555I			Unit
		Min	Typ	Max	Min	Typ	Max	
Supply voltage	V_{CC}	4.5	—	18	4.5	—	16	V
Supply current ($R_L = \infty$) - Note 1 Low state $V_{CC} = +5$ V $V_{CC} = +15$ V High state $V_{CC} = +5$ V	I_{CC} 	 — — —	 3 10 2	 5 12 —	 — — —	 3 10 2	 6 15 —	mA
Timing error (monostable) ($R_A = 1$ to 100 kΩ, C = 0.1 μF) — Initial accuracy (Note 2) — Drift with temperature — Drift with supply voltage	— 	 — — —	 0.5 30 0.05	 2 100 0.2	 — — —	 1 50 0.1	 — — —	 % ppm/°C %/V
Timing error (astable) (R_A, $R_B = 1$ kΩ to 100 kΩ, C = 0.1 μF, $V_{CC} = +15$ V) — Initial accuracy (Note 2) — Drift with temperature — Drift with supply voltage	— 	 — — —	 1.5 90 0.15	 — — —	 — — —	 2.25 150 0.3	 — — —	 % ppm/°C %/V
Control voltage level $V_{CC} = +15$ V $V_{CC} = +5$ V	V_{CL} 	 9.6 2.9	 10 3.33	 10.4 3.8	 9 2.6	 10 3.33	 11 4	V
Threshold voltage $V_{CC} = +15$ V $V_{CC} = +5$ V	V_{th} 	 9.4 2.7	 10 3.33	 10.6 4	 8.8 2.4	 10 3.33	 11.2 4.2	V
Threshold current - (Note 3)	I_{th}	—	0.1	0.25	—	0.1	0.25	μA
Trigger voltage $V_{CC} = +15$ V $V_{CC} = +5$ V	V_{trig} 	 4.8 1.45	 5 1.67	 5.2 1.9	 4.5 1.1	 5 1.67	 5.5 2.2	V
Trigger current ($V_{trig} = 0$ V)	I_{trig}	—	0.5	0.9	—	0.5	2.0	μA
Reset voltage - (Note 4)	V_{reset}	0.4	0.7	1	0.4	0.7	1	V
Reset current $V_{reset} = +0.4$ V $V_{reset} = 0$ V	I_{reset} 	 — —	 0.1 0.4	 0.4 1	 — —	 0.1 0.4	 0.4 1.5	mA
Low level output voltage $V_{CC} = +15$ V, $I_{O(sink)} = 10$ mA $I_{O(sink)} = 50$ mA $I_{O(sink)} = 100$ mA $I_{O(sink)} = 200$ mA $V_{CC} = +5$ V, $I_{O(sink)} = 8$ mA $I_{O(sink)} = 5$ mA	V_{OL} 	 — — — — — —	 0.1 0.4 2.0 2.5 0.1 0.05	 0.15 0.5 2.2 — 0.25 0.2	 — — — — — —	 0.1 0.4 2.0 2.5 0.3 0.25	 0.25 0.75 2.5 — 0.4 0.35	V
High level output voltage $V_{CC} = +15$ V, $I_{O(source)} = 200$ mA $I_{O(source)} = 100$ mA $V_{CC} = +5$ V, $I_{O(source)} = 100$ mA	V_{OH} 	 — 13.0 3	 12.5 13.3 3.3	 — — —	 — 12.75 2.75	 12.5 13.3 3.3	 — — —	V
Discharge pin leakage current (Output high)	$I_{dis(off)}$	—	1	100	—	1	100	nA
Discharge pin saturation voltage (Output low) - Note 5 $V_{CC} = +15$ V, $I_{dis} = 15$ mA $V_{CC} = +4.5$ V, $I_{dis} = 4.5$ mA	$V_{dis(sat)}$ 	 — —	 150 70	 — 100	 — —	 180 80	 — 200	mV
Output rise time Output fall time	t_r t_f	— —	100 100	200 200	— —	100 100	300 300	ns

Note 1 : Supply current when output is high is typically 1 mA less.

Note 2 : Tested at $V_{CC} = +5$ V and $V_{CC} = +15$ V.

Note 3 : This will determine the maximum value of $R_A + R_B$ for $+15$ V operation, the max total is $R = 20$ MΩ.

Note 4 : Specified with trigger input high.

Note 5 : No protection against excessive pin 7 current is necessary, providing the package dissipation rating will not be exceeded.

(TH)

NE555, SE555

MINIMUM PULSE WIDTH REQUIRED
FOR TRIGGERING

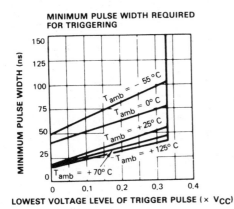

SUPPLY CURRENT vs SUPPLY VOLTAGE

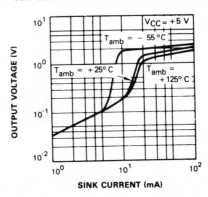

HIGH OUTPUT VOLTAGE vs OUTPUT SINK CURRENT

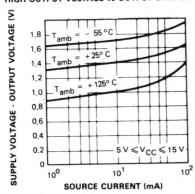

LOW OUTPUT VOLTAGE vs OUTPUT SINK CURRENT

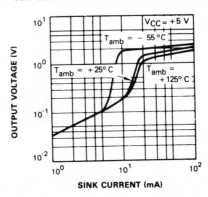

LOW OUTPUT VOLTAGE vs OUTPUT SOURCE CURRENT

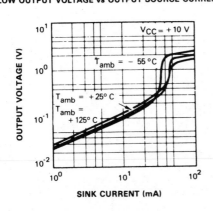

LOW OUTPUT VOLTAGE vs OUTPUT SINK CURRENT

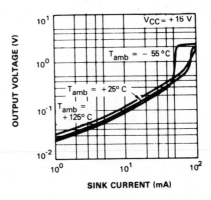

(TH)

NE555, SE555

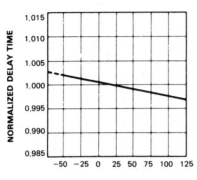

DELAY TIME vs TEMPERATURE

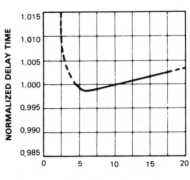

DELAY TIME vs SUPPLY VOLTAGE

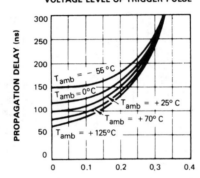

PROPAGATION DELAY vs
VOLTAGE LEVEL OF TRIGGER PULSE

(TH)

NE555, SE555

TYPICAL CHARACTERISTICS

MONOSTABLE OPERATION

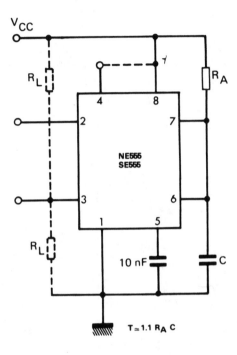

$T \simeq 1.1\, R_A\, C$

ASTABLE OPERATION

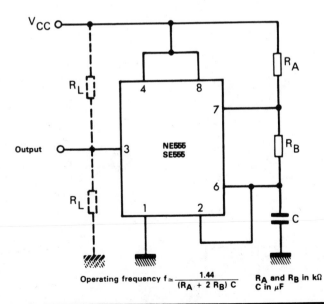

Operating frequency $f \simeq \dfrac{1.44}{(R_A + 2\,R_B)\,C}$

R_A and R_B in kΩ
C in μF

(TH)

NE555, SE555

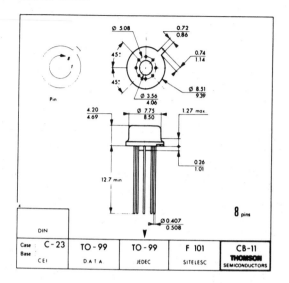

DIN				
Case : C-23	TO-99	TO-99	F 101	CB-11
Base				**THOMSON**
CEI	DATA	JEDEC	SITELESC	SEMICONDUCTORS

CB-11
(TO-99)

H SUFFIX
METAL CAN

CB-98

DP SUFFIX
PLASTIC PACKAGE
DG SUFFIX
CERDIP PACKAGE

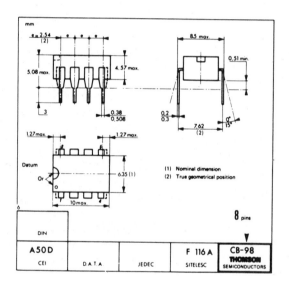

(1) Nominal dimension
(2) True geometrical position

DIN				
A50D			F 116 A	CB-98
				THOMSON
CEI	D.A.T.A.	JEDEC	SITELESC	SEMICONDUCTORS

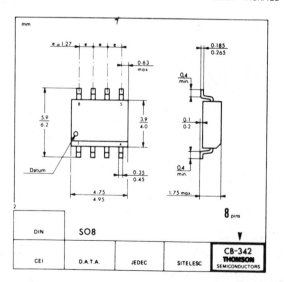

DIN	SO8			
				CB-342
CEI	D.A.T.A.	JEDEC	SITELESC	**THOMSON** SEMICONDUCTORS

CB-342

FP SUFFIX
PLASTIC
MICROPACKAGE

556 DUAL TIMERS

The NE556/SE556 dual timing circuits are highly stable controllers capable of producing accurate time delays, or oscillation.

The NE556/SE556 are dual 555. The two timers operate independently of each other sharing only Vcc and ground.

For astable operation as an oscillator, the free running frequency and the duty cycle are both accurately controlled with two external resistors and one capacitor.

The circuit may be triggered and reset on falling waveforms, and the output structure can source or sink up to 200 mA or drive TTL circuits.

- □ Replaces two NE555/SE555 timers.
- □ Timing from microseconds through hours.
- □ Operates in both astable and monostable modes.
- □ Adjustable duty cycle.
- □ High current output can source or sink 200 mA.
- □ Temperature stability of 0.005% per °C.

NE556, SE556

PIN ASSIGNMENT

(Top view)

ORDERING INFORMATION

Hi-Rel versions available - See chapter 14

PART NUMBER	TEMPERATURE RANGE	PACKAGE		
		DP	DG	FP
NE556C	0°C to + 70°C	●	●	●
SE556M	−55°C to +125°C		●	
Examples : NE556CDP, SE556MDG				

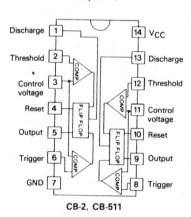

CB-2, CB-511

Fig. 3-19

(TH)

NE556, SE556

MAXIMUM RATINGS

Rating		Symbol	Value	Unit
Power supply voltage		V_{CC}	+18	V
Output current		I_O	200	mA
Power dissipation		P_{tot}	600	mW
Operating ambient temperature range	SE556	T_{oper}	−55 to +125	°C
	NE556		0 to +70	
Storage temperature range		T_{stg}	−65 to +150	°C

SCHEMATIC DIAGRAM (1/2 NE556)

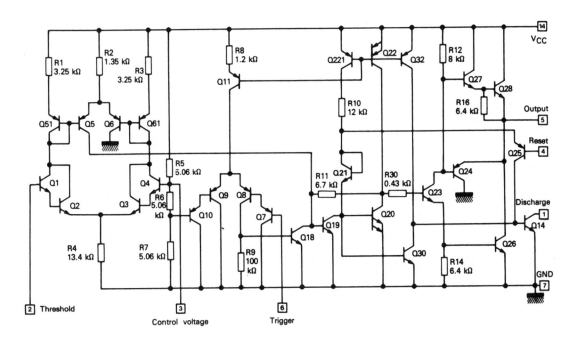

CASE	Discharge	Threshold	Control voltage	Reset	Outputs	Trigger	GND	V_{CC}
CB-2 CB-511	1, 13	2, 12	3, 11	4, 10	5, 9	6, 8	7	14

Fig. 3-19. Continued through page 143.

NE556, SE556

ELECTRICAL CHARACTERISTICS

$T_{amb} = +25°C$, $V_{CC} = +5$ to $+15$ V
(Unless otherwise specified)

Characteristic	Symbol	SE556			NE556			Unit
		Min	Typ	Max	Min	Typ	Max	
Supply voltage	V_{CC}	4.5	—	18	4.5	—	16	V

Supply current ($R_L = \infty$) - Note 1
  Low state, $V_{CC} = +5$ V
          $V_{CC} = +15$ V
  High state, $V_{CC} = +5$ V
	I_{CC}	
		—
		—
		—
			
			6
			20
			4
				
				10
				24
				—
					
					—
					—
					—
						
						6
						20
						4
							
							12
							30
							—
								mA

Timing error (monostable) $R_A = 1$ kΩ to 100 kΩ, C=0.1 μF — Initial accuracy (Note 2) — Drift with temperature — Drift with supply voltage	—	 — — —	 0.5 30 0.05	 1.5 100 0.2	 — — —	 0.75 50 0.1	 — — —	 % ppm/°C %/V
Timing error (astable) R_A, $R_B = 1$ kΩ to 100 kΩ, C=0.1 μF, $V_{CC} = +15$ V — Initial accuracy (Note 2) — Drift with temperature — Drift with supply voltage		 — — —	 1.5 90 0.15	 — — —	 — — —	 2.25 150 0.3	 — — —	 % ppm/°C %/V
Control voltage level $V_{CC} = +15$ V $V_{CC} = +5$ V	V_{CL}	 9.6 2.9	 10 3.33	 10.4 3.8	 9 2.6	 10 3.33	 11 4	V
Threshold voltage $V_{CC} = +15$ V $V_{CC} = +5$ V	V_{th}	 9.4 2.7	 10 3.33	 10.6 4	 8.8 2.4	 10 3.33	 11.2 4.2	V
Threshold current (Note 3)	I_{th}	—	0.1	0.25	—	0.1	0.25	μA
Trigger voltage $V_{CC} = +15$ V $V_{CC} = +5$ V	V_{trig}	 4.8 1.45	 5 1.67	 5.2 1.9	 4.5 1.1	 5 1.67	 5.5 2.2	V
Trigger current ($V_{trig} = 0$ V)	I_{trig}	—	0.5	0.9	—	0.5	2	μA
Reset voltage (Note 4)	V_{reset}	0.4	0.7	1	0.4	0.7	1	V
Reset current $V_{reset} = 0.4$ V $V_{reset} = 0$ V	I_{reset}	 — —	 0.1 0.4	 0.4 1	 — —	 0.1 0.4	 0.4 1.5	mA

Low level output voltage
 $V_{CC} = +15$ V, $I_{O(sink)} = 10$ mA
           $I_{O(sink)} = 50$ mA
           $I_{O(sink)} = 100$ mA
           $I_{O(sink)} = 200$ mA
 $V_{CC} = +5$ V, $I_{O(sink)} = 8$ mA
           $I_{O(sink)} = 5$ mA
	V_{OL}	
		—
		—
		—
		—
		—
		—
			
			0.1
			0.4
			2
			2.5
			0.1
			0.05
				
				0.15
				0.5
				2.25
				—
				0.25
				0.2
					
					—
					—
					—
					—
					—
					—
						
						0.1
						0.4
						2
						2.5
						0.3
						0.25
							
							0.25
							0.75
							2.75
							—
							0.4
							0.35
								V

Output voltage drop (high state) $V_{CC} = +15$ V, $I_{O(source)} = 200$ mA $I_{O(source)} = 100$ mA $V_{CC} = +5$ V, $I_{O(source)} = 100$ mA	V_{OH}	 — 13 3	 12.5 13.3 3.3	 — — —	 — 12.75 2.75	 12.5 13.3 3.3	 — — —	V
Discharge leakage current (pins 1 and 13)	$I_{dis(off)}$	—	1	100	—	1	100	nA
Discharge saturation voltage (pins 1 and 13) - Note 6 $V_{CC} = +15$ V, $I_{(7)} = 15$ mA $V_{CC} = +4.5$ V, $I_{(7)} = 4.5$ mA	$V_{dis(sat)}$	 — —	 150 70	 — 100	 — —	 180 80	 — 200	nV
Rise time of output	t_r	—	100	200	—	100	300	ns
Fall time of output	t_f	—	100	200	—	100	300	ns
Matching characteristics (Note 5) Initial accuracy (Note 2) Drift with temperature Drift with supply voltage	—	 — — —	 0.5 ± 10 0.1	 1 — 0.2	 — — —	 1 ± 10 0.2	 2 — 0.5	 % ppm/°C %/V

Note 1 : Supply current when output high typically 1 mA less at $V_{CC} = +5$ V.
Note 2 : Tested at $V_{CC} = +5$ V and $V_{CC} = +15$ V.
Note 3 : This will determine the maximum value of $R_A + R_B$ for 15 V operation. The maximum total is $R = 20$ MΩ.
Note 4 : Specified with trigger input high.
Note 5 : Matching characteristics refer to the difference between performance characteristics of each timer section.
Note 6 : No protection against excessive pin 1, 13 current is necessary providing the package dissipation rating will not be exceeded.

(TH)

NE556, SE556

MINIMUM PULSE WIDTH REQUIRED FOR TRIGGERING

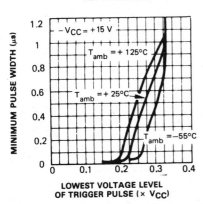

LOWEST VOLTAGE LEVEL
OF TRIGGER PULSE ($\times V_{CC}$)

SUPPLY CURRENT VERSUS SUPPLY VOLTAGE (each section)

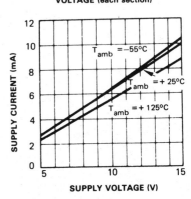

SUPPLY VOLTAGE (V)

HIGH OUTPUT VOLTAGE VERSUS OUTPUT SOURCE CURRENT

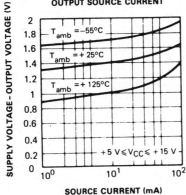

SOURCE CURRENT (mA)

LOW OUTPUT VOLTAGE VERSUS OUTPUT SINK CURRENT

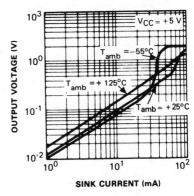

SINK CURRENT (mA)

LOW OUTPUT VOLTAGE VERSUS OUTPUT SINK CURRENT

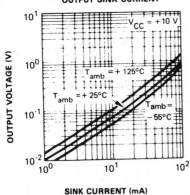

SINK CURRENT (mA)

LOW OUTPUT VOLTAGE VERSUS OUTPUT SINK CURRENT

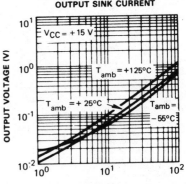

SINK CURRENT (mA)

(TH)

NE556, SE556

OUTPUT PROPAGATION DELAY VERSUS
VOLTAGE LEVEL OF TRIGGER PULSE

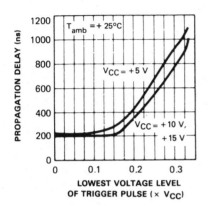

OUTPUT PROPAGATION DELAY VERSUS
VOLTAGE LEVEL OF TRIGGER PULSE

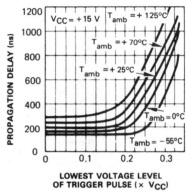

DISCHARGE TRANSISTOR (PINS 1, 13)
VOLTAGE VERSUS SINK CURRENT

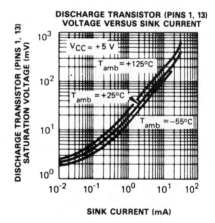

(TH)

NE556, SE556

50% DUTY CYCLE OSCILLATOR

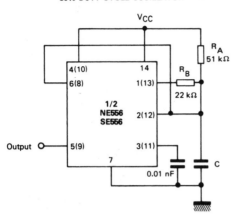

PULSE WIDTH MODULATOR

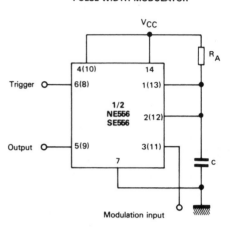

$$t1 = 0.693\ R_A.C$$

$$t2 = [(R_A\ R_B)\,/\,(R_A + R_B)]\,C \quad ln \quad \left[\frac{R_B - 2R_A}{2R_B - R_A}\right]$$

$$f = \frac{1}{t1 + t2} \qquad R_B < \frac{1}{2}\,R_A$$

TONE BURST GENERATOR

For a tone burst generator the first timer is used as a monostable and determines the tone duration when triggered by a positive pulse at pin 6. The second timer is enabled by the high output of the monostable. It is connected as an astable and determines the frequency of the tone.

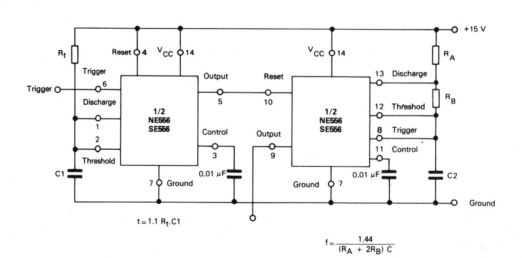

$$t = 1.1\ R_t.C1$$

$$f = \frac{1.44}{(R_A + 2R_B)\ C}$$

(TH)

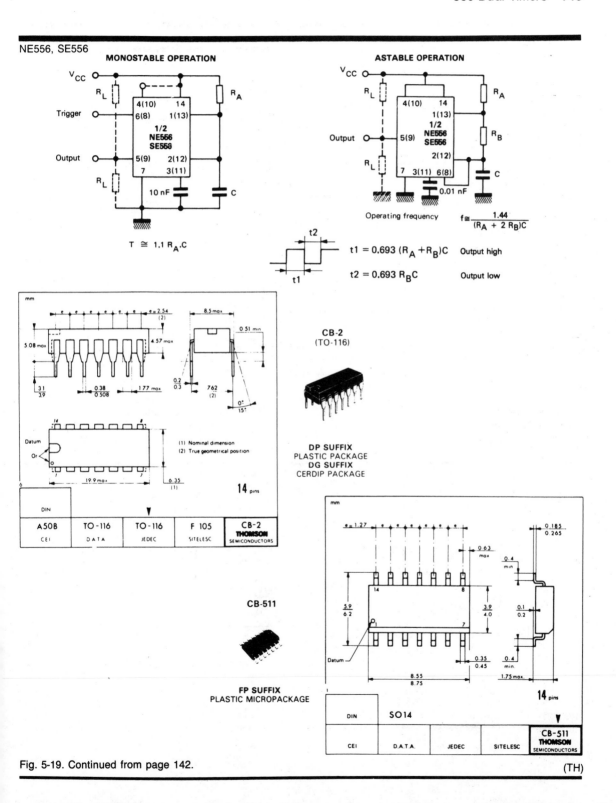

NE556, SE556

MONOSTABLE OPERATION

$T \cong 1.1\, R_A.C$

ASTABLE OPERATION

Operating frequency $\quad f \cong \dfrac{1.44}{(R_A + 2\,R_B)C}$

$t1 = 0.693\,(R_A + R_B)C \quad$ Output high

$t2 = 0.693\,R_B C \quad$ Output low

CB-2
(TO-116)

DP SUFFIX
PLASTIC PACKAGE
DG SUFFIX
CERDIP PACKAGE

CB-511

FP SUFFIX
PLASTIC MICROPACKAGE

SO14

Fig. 5-19. Continued from page 142.

(TH)

ICM7242

● **Process Control**
● **Machine Automation**

● Electro—pneumatic Drivers
● Multi—operation (Serial or Parallel Controlling)

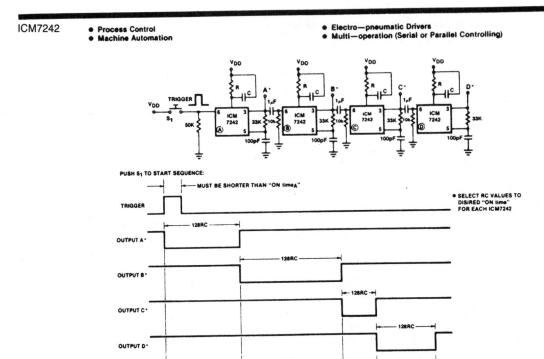

● **Process Control**
● **Machine Automation**

● **Electro-Pneumatic Drivers**
● **Multi-Operation (Serial or Parallel Controlling)**

Fig. 3-20   Sequence timer   (IN)

ICM7242

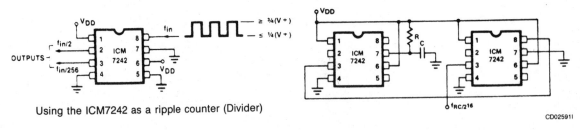

Using the ICM7242 as a ripple counter (Divider)

Low frequency reference (oscillator)

Fig. 3-21   (IN)

ICM7242

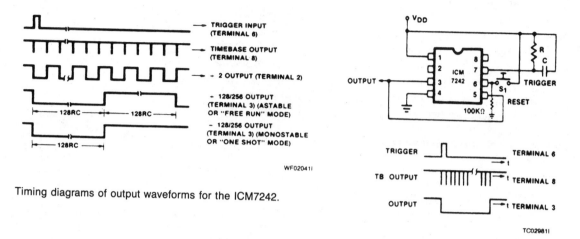

Timing diagrams of output waveforms for the ICM7242.

Monostable operation

Fig. 3-22

(IN)

ICM7242

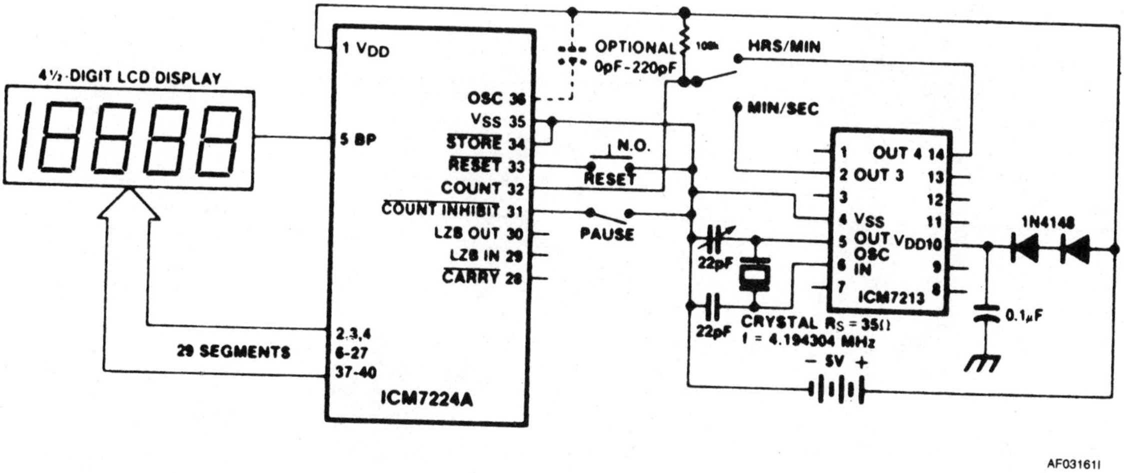

Fig. 3-23

Two-hour precision timer

(IN)

FAST RESET DC TIMER

Turning off a dc load after a time period is accomplished simply, with VP1220's 100 nA max gate current and large safe operating area. The timer responds with – 15 volts gate drive and SW 1 on. C1 together with R3 control the rate of gate voltage decay, and the power-on period. The capacitor discharges when D1 (low leakage type) conducts as SW1 is turned off during an interval period.

The addition of a linear temperature sensitive resistive component with R1, will increase accuracy over a wide range of operating conditions.

VP1220N1

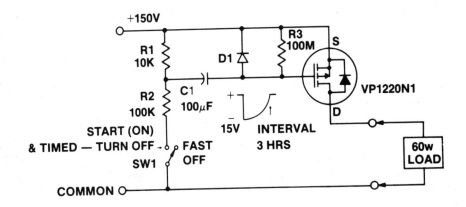

Fig. 3-24   (SU)

CMOS PRECISION PROGRAMMABLE 0-99 SECONDS/MINUTES LABORATORY TIMER

The ICM7250 is well suited as a laboratory timer to alert personnel of the expiration of a preselected interval of time.

When connected as shown in Fig. 3-25, the timer can accurately measure preselected time intervals of 0-99 seconds or 0-99 minutes. A 5 volt buzzer alerts the operator when the preselected time interval is over.

The circuit operates as follows: The time base is first selected with S1 (seconds or minutes), then units 0-99 are selected on the two thumbwheel switches S4 and S5. Finally, switch S2 is depressed to start the timer. Simultaneously the quartz crystal controlled divider circuits are reset, the ICM7250 is triggered and counting begins. The ICM7250 counts until the preprogrammed value is reached, whereupon it is reset, pin 10 of the CD4082B is enabled and the buzzer is turned on. Pressing S3 turns the buzzer off.

ICM7240, ICM7250, ICM7260

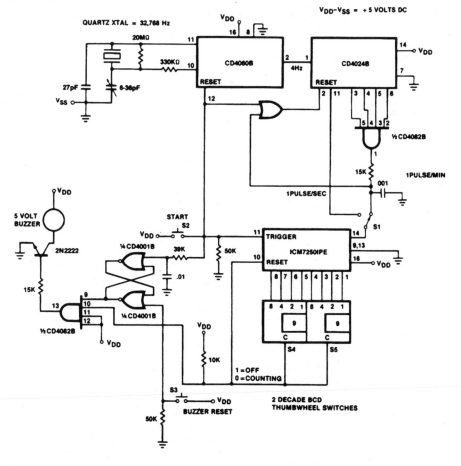

Fig. 3-25

(IN)

LOW POWER MICROPROCESSOR PROGRAMMABLE INTERVAL TIMER

The ICM7240 CMOS programmable binary timer can be configured as a low cost microprocessor controlled interval timer with the addition of a few inexpensive CD4000 series devices.

With the devices connected as shown in Fig. 3-26, the sequence of operation is as follows:

The microprocessor sends out an 8 bit binary code on its 8 bit I/O bus (the binary value needed to program the ICM7240), followed by four $\overline{\text{WRITE}}$ pulses into the CD417B decade counter. The first pulse resets the 8 bit latch, the second strobes the binary value into the 8 bit latch, the third triggers the ICM7240 to begin its timing cycle and the fourth resets the decade counter.

The ICM7240 then counts the interval of time determined by the R-C value on pin 13, and the programmed binary count on pins 1 through 8. At the end of the programmed time interval, the interrupt one-shot is triggered, informing the microprocessor that the programmed time interval is over.

With a resistor of approximately 10 MΩ and capacitor of 0.1 μF, the time base of the ICM7240 is one second. Thus, a time of 1-255 seconds can be programmed by the microprocessor, and by varying R or C, longer or shorter time bases can be selected.

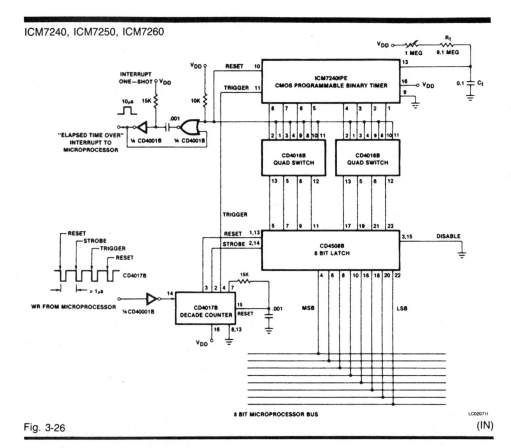

Fig. 3-26

(IN)

ICM7226 A/B

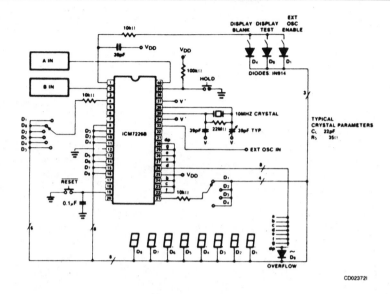

Fig. 3-27   10 MHz universal counter   (IN)

ICM7226 A/B

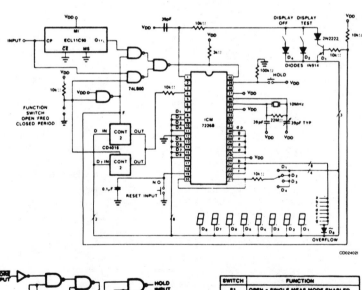

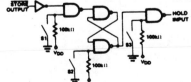

SWITCH	FUNCTION
S1	OPEN - SINGLE MEAS MODE ENABLED
S2	CLOSED - INITIATE NEW MEASUREMENT
S3	CLOSED - HOLD INPUT

SWITCH	FUNCTION
S1	OPEN-SINGLE MEAS MODE ENABLED
S2	CLOSED-INITIATE NEW MEASUREMENT
S3	CLOSED-HOLD INPUT

Fig. 3-28   100 MHz frequency, period counter   (IN)

ICM7226A/B

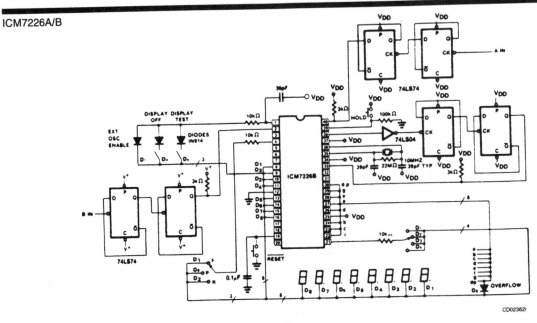

Notes: 1) If a 2.5MHz crystal is used, diode D1 and I.C.'s 1 and 2 can be eliminated.

40 MHz frequency, period counter

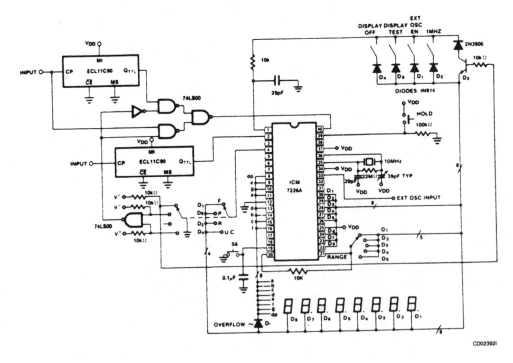

Fig. 3-29

100 MHz multi function counter

(IN)

COUNTERS

Figure 3-30 shows the schematic of an extremely simple unit counter that can be used for remote traffic counting, to name one application. The power cell stack should consist of 3 or 4 nickel cadmium rechargeable cells (nominal 3.6 or 4.8 volts). If 4 × 1.5 volt cells are used it is recommended that a diode be placed in series with the stack to guarantee that the supply voltage does not exceed 6 volts.

The input switch is shown to be a single pole double throw switch (SPDT). A single pole single throw switch (SPST) could also be used (with a pullup resistor), however, anti-bounce circuitry must be included in series with the counter input. In order to avoid contact bounce problems due to the SPDT switch the ICM7208 contains an input latch on chip.

The unit counter updates the display for each negative transition of the input signal. The information on the display will count, after reset, from 00 to 9,999,999 and then reset to 0000000 and begin to count up again. To blank leading zeroes, actuate reset at the beginning of a count. Leading zero blanking affects two digits at a time.

For battery operated systems the display may be switched off to conserve power.

Frequency Counter

The ICM7208 may be used as a frequency counter when used with an external frequency reference and gating logic. This can be achieved using the ICM7207 Oscillator Controller (Fig. 3-31). The ICM7207 uses a crystal controlled oscillator to provide the store and reset pulses together with the counting window. Figure 3-32 shows the recommended input gating waveforms to the ICM7208. At the end of a counting period (50% duty cycle) the counter input is inhibited. The counter information is then transferred and stored in latches, and can be displayed. Immediately after this information is stored, the counters are cleared and are ready to start a new count when the counter input is enabled.

Using a 6.5536 MHz quartz crystal and the ICM7207 driving the ICM7208, two ranges of counting may be obtained, using either 0.01 sec or 0.1 sec counter enable windows.

Previous comments on leading zero blanking, etc., apply as per the unit counter.

The ICM7207 provides the multiplex frequency reference of 1.6 kHz.

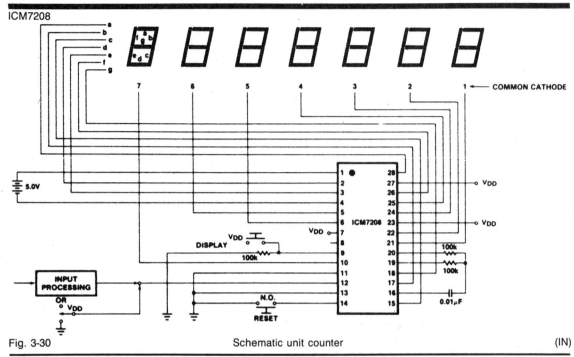

Fig. 3-30   Schematic unit counter   (IN)

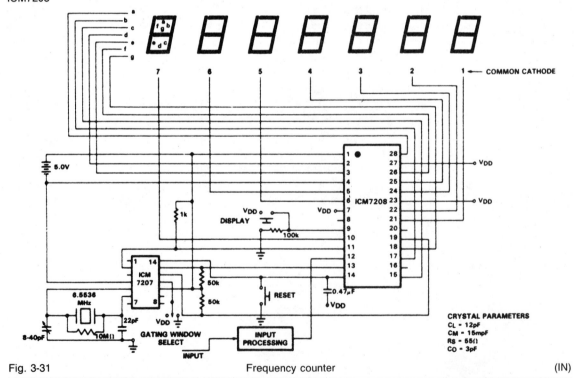

Fig. 3-31   Frequency counter   (IN)

ICM7208

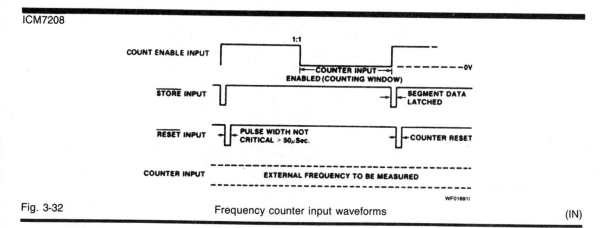

Fig. 3-32   Frequency counter input waveforms   (IN)

ICM7249

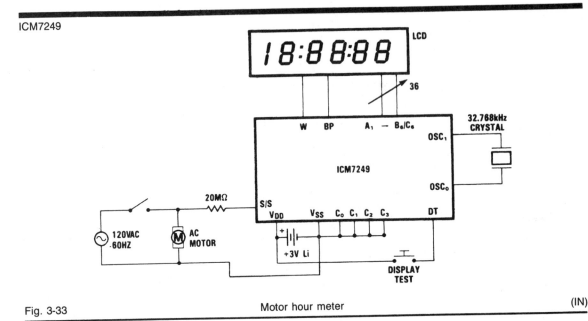

Fig. 3-33   Motor hour meter   (IN)

ICM7224/ICM7225

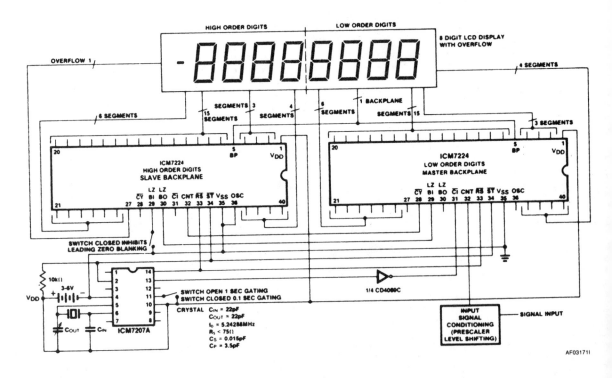

Fig. 3-34   Eight-digit precision frequency counter   (IN)

ICM7216 A/B/C/D

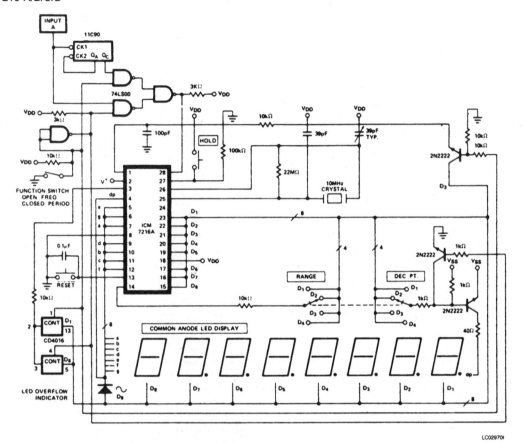

100 MHz frequency, 2 MHz period counter

LC029701

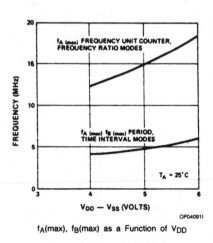

OP040911

$f_A(max)$, $f_B(max)$ as a Function of V_{DD}

Fig. 3-35   Typical operating characteristics   (IN)

ICM7216 A/B/C/D

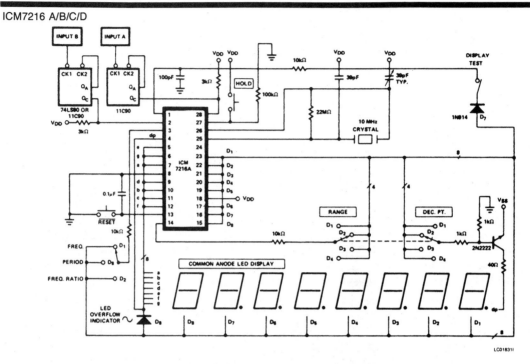

Fig. 3-36   100 MHz multifunction counter   (IN)

ICM7216 A/B/C/D

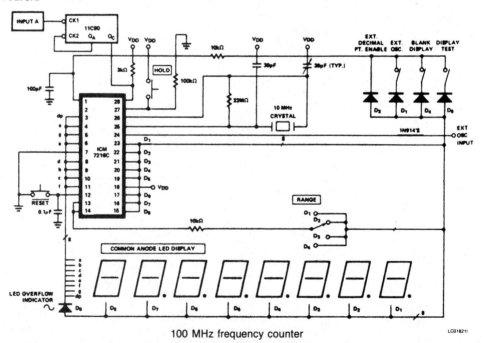

Fig. 3-37   100 MHz frequency counter   (IN)

ICM7216 A/B/C/D

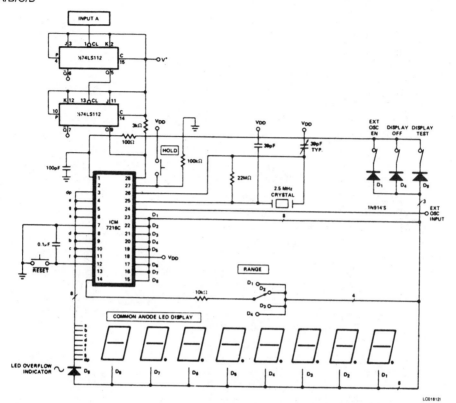

Fig. 3-38   40 MHz frequency counter   (IN)

ICM7216 A/B/C/D

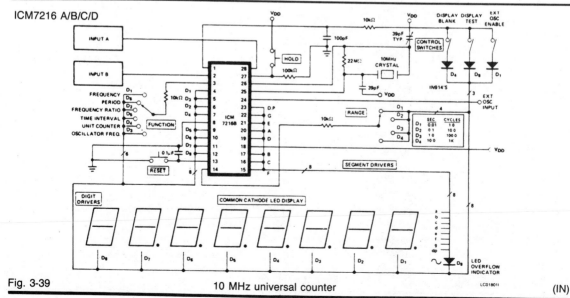

Fig. 3-39   10 MHz universal counter   (IN)

ICM7215

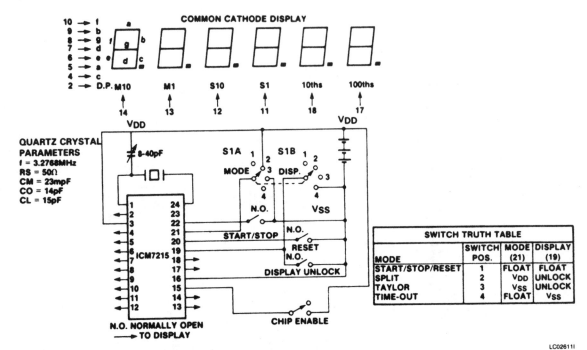

SWITCH TRUTH TABLE			
MODE	SWITCH POS.	MODE (21)	DISPLAY (19)
START/STOP/RESET	1	FLOAT	FLOAT
SPLIT	2	VDD	UNLOCK
TAYLOR	3	VSS	UNLOCK
TIME-OUT	4	FLOAT	VSS

Fig. 3-40   Stopwatch circuit   (IN)

ICM7245

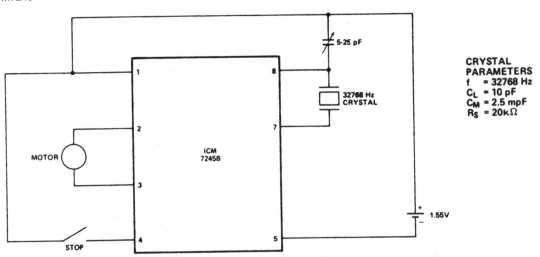

CRYSTAL PARAMETERS
f = 32768 Hz
C_L = 10 pF
C_M = 2.5 mpF
R_S = 20kΩ

Fig. 3-41   Typical watch circuit   (IN)

FOUR QUADRANT ANALOG MULTIPLIER

The ICL8013 is a four quadrant analog multiplier whose output is proportional to the algebraic product of two input signals. Feedback around an internal op-amp provides level shifting and can be used to generate division and square root functions. A simple arrangement of potentiometers may be used to trim gain accuracy, offset voltage and feedthrough performance. The high accuracy, wide bandwidth, and increased versatility of the ICL8013 make it ideal for all multiplier applications in control and instrumentation systems. Applications include rms measuring equipment, frequency doublers, balanced modulators and demodulators, function generators, and voltage controlled amplifiers.

Features

□ Accuracy of $\pm 0.5\%$ ("A" Version)
□ Full ± 10 V Input Voltage Range
□ 1 MHz Bandwidth
□ Uses Standard ± 15 V Supplies
□ Built-In Op Amp Provides Level Shifting, Division and Square Root Functions

Multiplication

In the standard multiplier connection, the Z terminal is connected to the op amp output. All of the modulator output current thus flows through the feedback resistor R27 and produces a proportional output voltage.

Multiplier Trimming Procedure

□ Set $X_{IN} = Y_{IN} = 0$ V and adjust Z_{OS} for zero Output.
□ Apply a ± 10 V low frequency (≤ 100 Hz) sweep (sine or triangle) to Y_{IN} with $X_{IN} = 0$ V, and adjust X_{OS} for minimum output.
□ Apply the sweep signal to X_{IN} with $Y_{IN} = 0$ V and adjust Y_{OS} for minimum Output.
□ Readjust Z_{OS} if necessary.
□ With $X_{IN} = 10.0$ Vdc and the sweep signal applied to Y_{IN}, adjust the Gain potentiometer for Output = Y_{IN}. This is easily accomplished with a differential scope plug-in (A + B) by inverting one signal and adjusting Gain control for (Output – Y_{IN}) = Zero.

Division

If the Z terminal is used as an input, and the output of the op-amp connected to the Y input, the device functions as a divider. Since the input to the op-amp is at virtual ground, and requires negligible bias current, the overall feedback forces the modulator output current to equal the current produced by Z.

$$\text{Therefore } I_O = X_{IN} \bullet Y_{IN} = \frac{Z_{IN}}{R} = 10\, Z_{IN}$$

$$\text{Since } Y_{IN} = E_{OUT}, \quad E_{OUT} = \frac{10 \, Z_{IN}}{X_{IN}}$$

Note that when connected as a divider, the X input must be a negative voltage to maintain overall negative feedback.

Divider Trimming Procedure

☐ Set trimming potentiometers at mid-scale by adjusting voltage on pins 7, 9 and 10 (X_{OS}, Y_{OS}, Z_{OS}) for zero volts.
☐ With Z_{IN} = O V, trim Z_{OS} to hold the Output constant, as X_{IN} is varied from – 10 V through – 1 V.
☐ With Z_{IN} = 0 V and X_{IN} = – 10.0 V adjust Y_{OS} for zero Output voltage.
☐ With Z_{IN} = X_{IN} (and/or Z_{IN} = – X_{IN}) adjust X_{OS} for minimum worst-case variation of Output, as X_{IN} is varied from – 10 V to – 1 V.
☐ If the last procedure required a large initial adjustment repeat the first two processes of these procedures.
☐ With Z_{IN} = X_{IN} (and/or Z_{IN} = – X_{IN}) adjust the gain control until the output is the closest average around + 10.0 V (– 10 V for Z_{IN} = – X_{IN}) as X_{IN} is varied from – 10 V to – 3 V.

Squaring

The squaring function is achieved by simply multiplying with the two inputs tied together. The squaring circuit may also be used as the basis for a frequency doubler since $\cos^2 \omega t = \frac{1}{2} (\cos 2\omega t + 1)$.

Square Root

Tying the X and Y inputs together and using overall feedback from the op amp results in the square root function. The output of the modulator is again forced to equal the current produced by the Z input.

$$I_O = X_{IN} \bullet Y_{IN} = (-E_{OUT})^2 = 10 \, Z_{IN}$$

$$E_{OUT} = -\sqrt{10 \, Z_{IN}}$$

The output is a negative voltage which maintains overall negative feedback. A diode in series with the op amp output prevents the latchup that would otherwise occur for negative input voltages.

Square Root Trimming Procedure

☐ Connect the ICL8013 in the Divider configuration.
☐ Adjust Z_{OS}, Y_{OS}, X_{OS}, and Gain using the first six procedures of Divider Trimming Procedure.
☐ Convert to the square root configuration by connecting X_{IN} to the output and inserting a diode between pin 4 and the output node.
☐ With Z_{IN} = 0 V adjust Z_{OS} for zero output voltage.

ICL8013

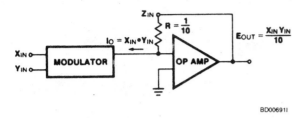

Multiplier block diagram

BD006911

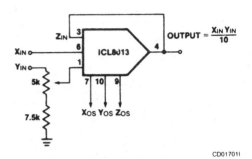

CD017011

Actual circuit connection

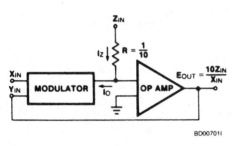

BD007011

Division block diagram

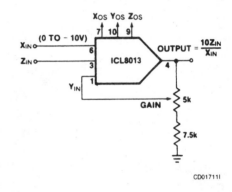

CD017111

Actual circuit connection

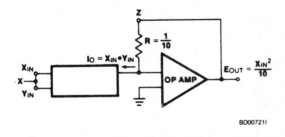

BD007211

Square block diagram

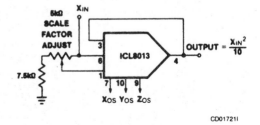

CD017211

Actual circuit connection

Fig. 3-42

ICL8013

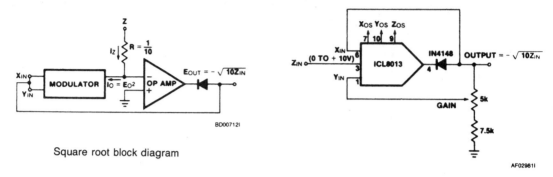

Square root block diagram

Square root

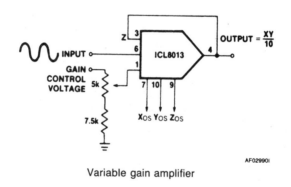

Variable gain amplifier

Fig. 3-43   (IN)

Variable Gain Amplifier

Most applications for the ICL8013 are straightforward variations of the simple arithmetic functions described above. Although the circuit description frequently disguises the fact, it has already been shown that the frequency doubler is nothing more than a squaring circuit. Similarly the variable gain amplifier is nothing more than a multiplier, with the input signal applied at the X input and the control voltage applied at the Y input.

Definition of Terms

Multiplication/Division Error: This is the basic accuracy specification. It includes terms due to linearity, gain, and offset errors, and is expressed as a percentage of the full scale output.

Feedthrough: With either input at zero, the output of an ideal multiplier should be zero regardless of the signal applied to the other input. The output seen in a non-ideal multiplier is known as the feedthrough.

Nonlinearity: The maximum deviation from the best straight line constructed through the output data, expressed as a percentage of full scale. One input is held constant and the other swept through its nominal range. The nonlinearity is the component of the total multiplication/division error which cannot be trimmed out.

LOW COST MULTIFUNCTION CONVERTER

Burr-Brown's multifunction converter model 4302 is a low cost solution to many analog conversion needs. Much more than just another multiplier/divider, the 4302 out performs many analog circuit functions with a very high degree of accuracy at a very low total cost to the user.

General specifications for the Model 4302 Multifunction Converter are presented on this page. These specifications characterize the 4302 as a versatile three input multifunction converter.

The following pages are applications oriented to help you apply the 4302 to your particular circuit function need. These pages contain dedicated circuit configurations in order to produce the functions of: multiplication, division, exponentiation, square rooting, squaring, sine, cosine, arctangent, and vector algebra.

It is the purpose of this product data sheet to enable you to apply the 4302 to your analog conversion needs quickly and efficiently.

Many of the following circuit configurations using the 4302 require a reference voltage for scaling purposes. The reference voltage is shown to be + 15 Vdc (+ 15 Vdc ref.) since in most cases the + 15 Vdc power source for the 4302 has sufficient time and temperature related stability to achieve the specified typical accuracies.

If the particular supplies which are available for powering the 4302 do not have the necessary stability for the required conversion accuracy, an additional + 15 Vdc precision supply may be required.

Multiplier

In multiplier applications the 4302 provides high accuracy at a low cost. The 4302 accepts inputs up to + 10 Vdc and provides a typical accuracy of $\pm 0.25\%$ of full scale.

(1) Set R1 so that with $E_1 = E_2 = +10.00$ Vdc, $E_o = +10.00$ Vdc.

Divider

As a divider, the 4302 outperforms many of the multiplier/dividers on the market at a much lower cost. In the divider configuration the 4302 boasts a typical conversion accuracy of $\pm 0.25\%$ of full scale.

Notes:
(1) Set R1 so that with $E_1 = E_3 = +10.00$ Vdc, $E_o = +10.00$ Vdc.
(2) Set R2 so that with $E_1 = E_3 = +0.10$ Vdc, $E_o = +10.00$ Vdc.
(3) Set R3 so that with $E_1 = +0.01$ Vdc and with $E_3 = +0.10$ Vdc, $E_o = +1.00$ Vdc.
(4) Repeat steps 1 through 3 as necessary to achieve the specified output voltages.

Exponential Functions

Model 4302 may be used as exponentiator over a range of exponents from 0.2 to 5. The exponents 0.5 and 2, square rooting and squaring respectively, are

4302

FUNCTIONS	ACCURACY
MULTIPLY	±0.25%
DIVIDE	±0.25%
SQUARE	±0.03%
SQUARE ROOT	±0.07%
EXPONENTIATE	±0.15% (m = 5)
ROOTS	±0.2% (m = .2)
SINE θ	±0.5%
COSINE θ	±0.8%
TAN $^{-1}$ (Y/X)	±0.6%
$\sqrt{X^2 + Y^2}$	±0.07%

Fig. 3-44

(BB)

often used functions and are treated below. Other values of exponents (m) may be useful in terms of linearization of nonlinear functions or simply for producing the mathematical conversions. Characteristics of m = 0.2 and m = 5 are presented in Table 3-2. For other values of m the curves presented in Fig. 3-47 may be used to interpolate the error for a nonspecified value of m.

Notes:

(1) Connect a 100 Ω potentiometer as shown in Fig. 3-50 for either roots (0.2 $\leq$ m < 1) or powers (1 < m $\leq$ 5).

(2) Set R1 so that with E_1 = +10.00 Vdc, E_o = +10.00 Vdc.

(3) Select a +dc voltage level (E_1) such that the output voltage (E_o), as acted upon by the desired exponent, will not exceed +10.00 Vdc. A level which is mid-range for input values of interest is an appropriate one to use. Set R2 so that the output voltage (E_o) is the value expected for the chosen values of input (E_1) and exponent (m).

(4) Repeat steps (2) through (4) as necessary.

When taking roots of smaller input levels, a modified transfer equation [E_o = $(10E_1)^m$] will provide improved conversion accuracy. To achieve this transfer function: 1) apply a +1.5 Vdc ref in place of the +15 Vdc ref shown in Fig. 3-50, 2) make R3 a 1.40 MΩ resistor, and rearrange R1 and R3 as 1.5 Vdc ref and 3) follow all notes except in note (2) apply +0.10 Vdc to pin 7 to set R1 to E_o = +1.00 Vdc.

Table 3-2.

Transfer Function	$E_o = 10 \left(\dfrac{E_1}{10} \right)^m$
Total Conversion Error (typical)	
m = 0.2	
0.5 VDC < E_1 < 10 VDC	±2 m VDC
0.1 VDC < E_1 < 0.5 VDC	±25 m VDC
m = 5	
1.0 VDC < E_1 < 10 VDC	±15 m VDC
Exponent Range (continuous)	0.2 $\leq$ m $\leq$ 5
Input Voltage Range	0 to +10 VDC
Output Voltage Range	0 to +10 VDC

4302

ELECTRICAL

MODEL	4302
TRANSFER FUNCTION	$E_O = V_Y \left(\dfrac{V_Z}{V_X}\right)^m$
RATED OUTPUT	
Voltage	+10.0 V
Current	5 mA
INPUT	
Signal Range	$0 \leqslant (V_X, V_Y, V_Z) \leqslant +10$ V
Absolute Maximum	$(V_X, V_Y, V_Z) \leqslant \pm18$ V
Impedance (X/Y/Z)	100 kΩ/90 kΩ/100 kΩ
EXPONENT RANGE	
Roots ($0.2 \leqslant m < 1$)	$m = \dfrac{R_2}{R_1 + R_2}$ Refer to
	Functional
Powers ($1 < m \leqslant 5$)	$m = \dfrac{R_1 + R_2}{R_2}$ Diagram below
($m = 1$)	$R_1 = 0$ Ω, R_2 not used
POWER REQUIREMENTS	
Rated Supply	±15 VDC
Range	±12 to ±18 VDC
Quiescent Current	±10 mA
TEMPERATURE RANGE	
Operating	−25°C to +85°C
Storage	−25°C to +85°C

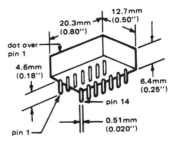

MECHANICAL

12.7mm (0.50")
20.3mm (0.80")
dot over pin 1
4.6mm (0.18")
6.4mm (0.25")
pin 14
0.51mm (0.020")
pin 1

Row Spacing: 7.6mm (0.300")
Weight: 3.4 grams (0.12 oz.)
Connector: 14-pin DIP
0145MC

Pin material and plating composition conform to Method 208 (solderability) of Mil-Std-202.

PIN CONNECTIONS

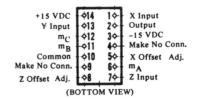

+15 VDC	14	1	X Input
Y Input	13	2	Output
m_C	12	3	−15 VDC
m_B	11	4	Make No Conn.
Common	10	5	X Offset Adj.
Make No Conn.	9	6	m_A
Z Offset Adj.	8	7	Z Input

(BOTTOM VIEW)

Fig. 3-45   (BB)

4302

4302 FUNCTIONAL DIAGRAM

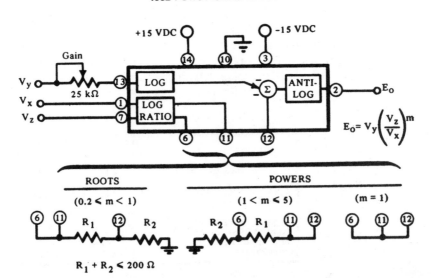

$$E_O = V_y \left(\frac{V_z}{V_x}\right)^m$$

$R_1 + R_2 \leqslant 200$ Ω

Fig. 3-46   (BB)

4302

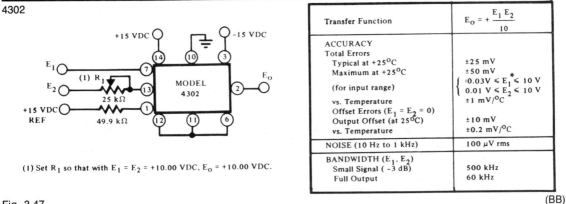

(1) Set R_1 so that with $E_1 = E_2 = +10.00$ VDC, $E_o = +10.00$ VDC.

Transfer Function	$E_o = + \dfrac{E_1 E_2}{10}$
ACCURACY Total Errors	
Typical at +25°C	±25 mV
Maximum at +25°C	±50 mV
(for input range)	$\begin{cases} 0.03V \le E_1{}^* \le 10\ V \\ 0.01\ V \le E_2 \le 10\ V \end{cases}$
vs. Temperature	±1 mV/°C
Offset Errors ($E_1 = E_2 = 0$)	
Output Offset (at 25°C)	±10 mV
vs. Temperature	±0.2 mV/°C
NOISE (10 Hz to 1 kHz)	100 µV rms
BANDWIDTH (E_1, E_2)	
Small Signal (-3 dB)	500 kHz
Full Output	60 kHz

Fig. 3-47   (BB)

4302

Transfer Function	$E_o = +10\ (E_1/E_3)$
ACCURACY Total Errors	
Typical at +25°C	±25 mV
Maximum at +25°C	±50 mV
(for $E_1 \le E_3$ and input range)	$\begin{cases} 0.03V \le E_1{}^* \le 10\ V \\ 0.1\ V \le E_3 \le 10\ V \end{cases}$
vs. Temperature	±1 mV/°C
Offset Errors ($E_1 = 0$, $E_3 = +10$ V)	
Output Offset (at 25°C)	±10 mV
vs. Temperature	±1 mV/°C
NOISE (10 Hz to 1 kHz)	
$E_3 = +10$ V	100 µV rms
$E_3 = +0.1$ V	300 µV rms
BANDWIDTH (E_1, E_3)	
Small Signal (-3 dB)	500 kHz
Full Output	
($E_3 = +10$ V)	60 kHz

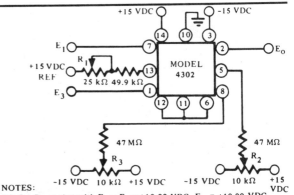

NOTES:
(1) Set R_1 so that with $E_1 = E_3 = +10.00$ VDC, $E_o = +10.00$ VDC.
(2) Set R_2 so that with $E_1 = E_3 = +0.10$ VDC, $E_o = +10.00$ VDC.
(3) Set R_3 so that with $E_1 = +0.01$ VDC and with $E_3 = +0.10$ VDC, $E_o = +1.00$ VDC.
(4) Repeat steps 1 through 3 as necessary to achieve the specified output voltages.

*The input voltage may be extended below 0.03V by connecting a 0.047 µF capacitor between pins 11 and 5, causing a slight reduction in bandwidth. (Multiply and Divide Modes).

Fig. 3-48   (BB)

4302

$$E_o = 10 \left(\frac{E_1}{10}\right)^m \quad m = 0.2$$

Fig. 3-49   **Exponentiator Transfer Characteristics**   (BB)

4302

Use these connections when taking roots of small input levels.

POWERS   ROOTS*

Fig. 3-50   (BB)

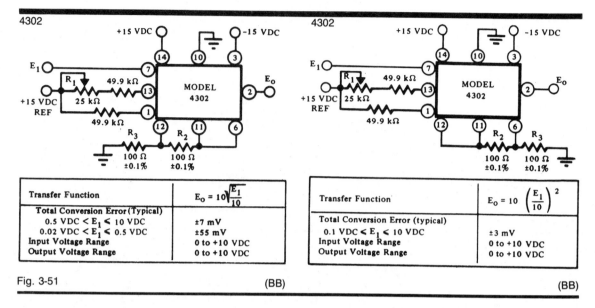

Transfer Function	$E_o = 10\sqrt{\dfrac{E_1}{10}}$
Total Conversion Error (Typical)	
$0.5\ \text{VDC} < E_1 \leqslant 10\ \text{VDC}$	$\pm 7\ \text{mV}$
$0.02\ \text{VDC} < E_1 \leqslant 0.5\ \text{VDC}$	$\pm 55\ \text{mV}$
Input Voltage Range	0 to +10 VDC
Output Voltage Range	0 to +10 VDC

Fig. 3-51   (BB)

Transfer Function	$E_o = 10\left(\dfrac{E_1}{10}\right)^2$
Total Conversion Error (typical)	
$0.1\ \text{VDC} \leqslant E_1 \leqslant 10\ \text{VDC}$	$\pm 3\ \text{mV}$
Input Voltage Range	0 to +10 VDC
Output Voltage Range	0 to +10 VDC

(BB)

Square Root

As a Square Rooter (m = 0.5), the 4302 provides a typical total conversion accuracy of ±0.07%. Refer to Fig. 3-51 and notes for connections and adjustments respectively.

Notes:

(1) Connect pins 12, 11 and 6 together. Set R1 such that with E_1 = +10.00 Vdc; E_o = +10.00 Vdc.

Fig. 3-52

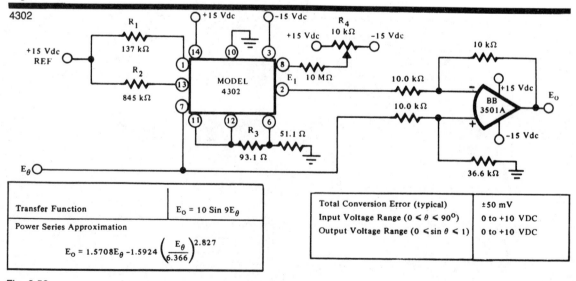

Transfer Function	$E_o = 10\ \text{Sin}\ 9E_\theta$
Power Series Approximation	
$E_o = 1.5708E_\theta - 1.5924\left(\dfrac{E_\theta}{6.366}\right)^{2.827}$	

Total Conversion Error (typical)	±50 mV
Input Voltage Range ($0 \leqslant \theta \leqslant 90^\circ$)	0 to +10 VDC
Output Voltage Range ($0 \leqslant \sin\theta \leqslant 1$)	0 to +10 VDC

Fig. 3-53   (BB)

4302

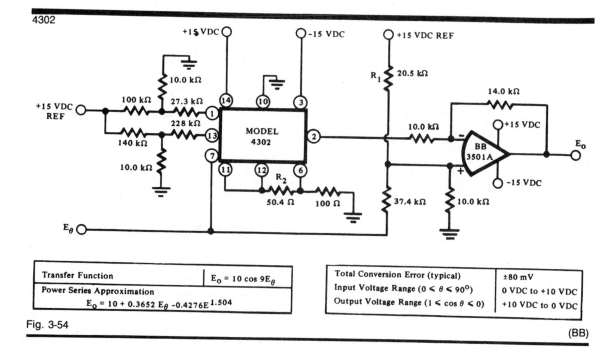

Transfer Function	$E_o = 10 \cos 9E_\theta$
Power Series Approximation	
$E_o = 10 + 0.3652\, E_\theta - 0.4276E^{1.504}$	

Total Conversion Error (typical)	±80 mV
Input Voltage Range (0 ≤ θ ≤ 90°)	0 VDC to +10 VDC
Output Voltage Range (1 ≤ cos θ ≤ 0)	+10 VDC to 0 VDC

Fig. 3-54

(BB)

(2) Connect 100 Ω resistors as shown in Fig. 3-52.
(3) For greater conversion accuracy, R2 & R3 may be replaced by a potentiometer as shown in Fig. 3-52.

Square

Configured as a Square Function Converter (m = 2), the 4302 produces high conversion accuracies of typically 0.03%. Please refer to Fig. 3-52 and accompanying notes.

Notes:
(1) Set R1 such that with $E_1 = +10.00$ Vdc, $E_o = +10.00$ Vdc.
(2) Connect 100 Ω resistors as shown in Fig. 3-52.
(3) For greater conversion accuracy R2 & R3 may be replaced by a potentiometer as shown in Fig. 3-52.

Arctangent

Model 4302 and the associated circuitry shown below will produce the inverse tangent of a ratio. This application is particularly well suited to conversion from rectangular coordinates to polar coordinates where

$$E\theta = \tan^{-1} \frac{E_y}{E_x}$$

4302

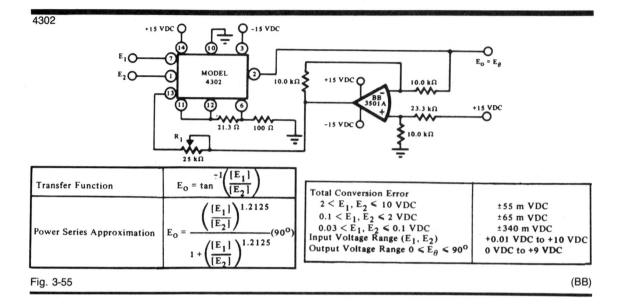

Transfer Function	$E_o = \tan^{-1}\left(\dfrac{|E_1|}{|E_2|}\right)$
Power Series Approximation	$E_o = \dfrac{\left(\dfrac{|E_1|}{|E_2|}\right)^{1.2125}}{1 + \left(\dfrac{|E_1|}{|E_2|}\right)^{1.2125}}(90^\circ)$

Total Conversion Error	
$2 < E_1, E_2 \leqslant 10$ VDC	± 55 m VDC
$0.1 < E_1, E_2 \leqslant 2$ VDC	± 65 m VDC
$0.03 < E_1, E_2 \leqslant 0.1$ VDC	± 340 m VDC
Input Voltage Range (E_1, E_2)	+0.01 VDC to +10 VDC
Output Voltage Range $0 \leqslant E_\theta \leqslant 90^\circ$	0 VDC to +9 VDC

Fig. 3-55   (BB)

The accuracy of conversion depends upon the levels of the input signals. Please refer to Table 3-2.

Note:
(1) Set R1 so that with $E_1 = E_2 = +10.00$ Vdc, $E_o = +4.500$ Vdc ± 1 mVdc.

Vector Magnitude Function

The model 4302 will produce the square root of the sum of the squares of two inputs. This function is companion to the arctangent of a ratio for the conversion of rectangular to polar coordinates.

Notes:
(1) Figure 3-56 shows one practical way to implement the transfer function $E_o = \sqrt{E_1{}^2 + E_2{}^2}$ using 4302. It shows use of model 3501A op amp. Model 3501's rated output is ± 10 V. This limits the range of E_1 and E_2, such that the conditions $E_1 \leq \sqrt{100 - E_2}$ and $|E_2| < -(5 - E_1{}^2/20)$ and $\sqrt{E_1{}^2 + E_2{}^2} \leq 10$ are always satisfied.
(a) The above conditions imply, $0\,V \leq E_1 \leq 10\,V$ and $-5\,V \leq E_2 \leq 5\,V$.
(b) The above conditions also imply that for applications where $E_1 = |E_2|$ the range would be limited to 4.142 V max.
(2) Use of model 3627 as shown in Fig. 3-57 would directly substitute the eight 10 kΩ resistors and the two model 3501 A op amps. This would reduce the number of components needed to implement vector magnitude function and reduce overall cost.

4302

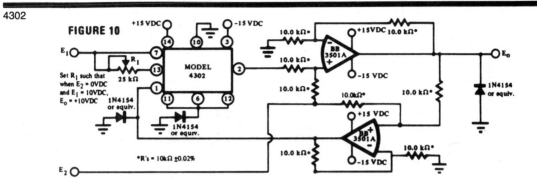

Transfer Function	$E_o = \sqrt{E_1{}^2 + E_2{}^2}$
Input Voltage Range E_1 $\quad\quad\quad\quad\quad\quad E_2$	0 to +10VDC $\quad$ −10VDC to +10VDC
(refer to notes 1 and 2)	
Output Voltage Range	0 to +10VDC
Conversion Error	±7m VDC

Fig. 3-56 (BB)

4302

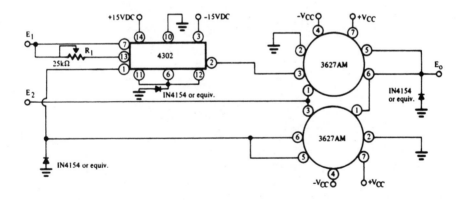

Fig. 3-57 (BB)

4302

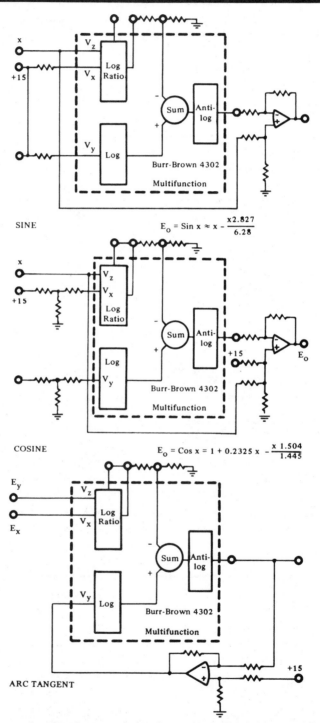

SINE

$$E_O = \text{Sin } x \approx x - \frac{x^{2.827}}{6.28}$$

COSINE

$$E_O = \text{Cos } x = 1 + 0.2325 x - \frac{x^{1.504}}{1.445}$$

ARC TANGENT

Fig. 3-58   Connections for trigonometric functions using multifunction converter.   (BB)

4
Interfacing Circuits

Typical A/D and D/A circuits   Typical D/A and A/D circuits (general configurations)

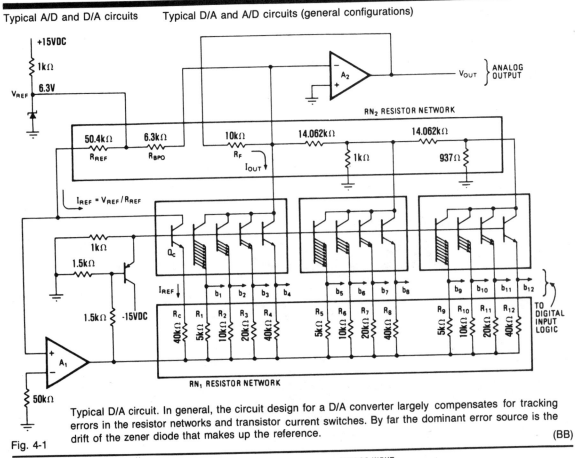

Typical D/A circuit. In general, the circuit design for a D/A converter largely compensates for tracking errors in the resistor networks and transistor current switches. By far the dominant error source is the drift of the zener diode that makes up the reference.   (BB)

Fig. 4-1

Typical A/D and D/A circuits

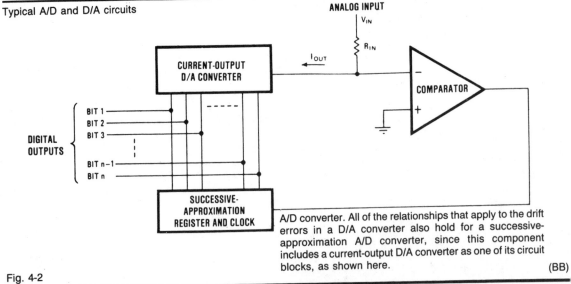

A/D converter. All of the relationships that apply to the drift errors in a D/A converter also hold for a successive-approximation A/D converter, since this component includes a current-output D/A converter as one of its circuit blocks, as shown here.   (BB)

Fig. 4-2

ADC10HT

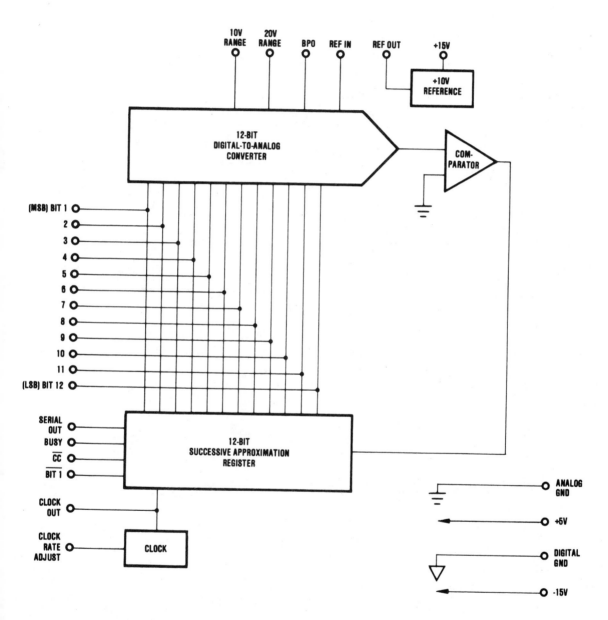

The ADC10HT 12-bit hybrid A/D converter.

Fig. 4-3

(BB)

LM118, LM218, LM318

D/A CONVERTER USING LADDER NETWORK

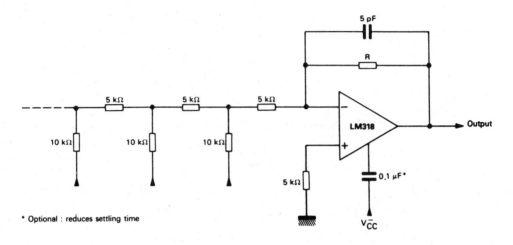

* Optional : reduces settling time

Fig. 4-4   (TH)

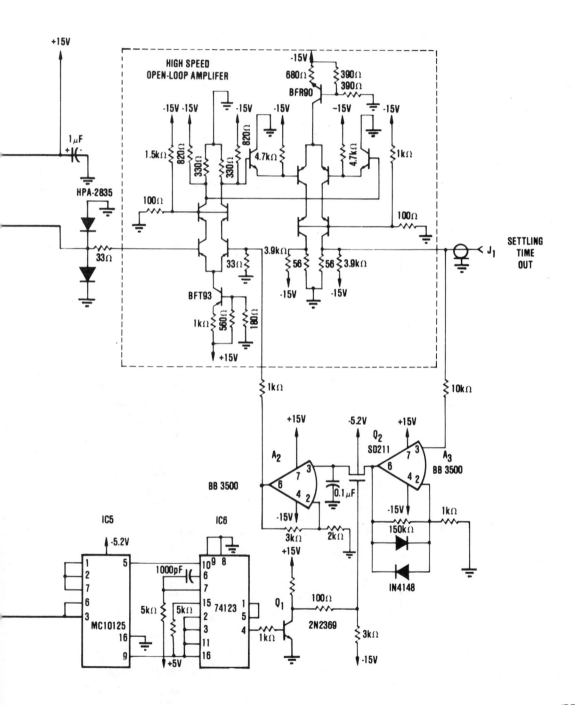

(BB)

DAC63

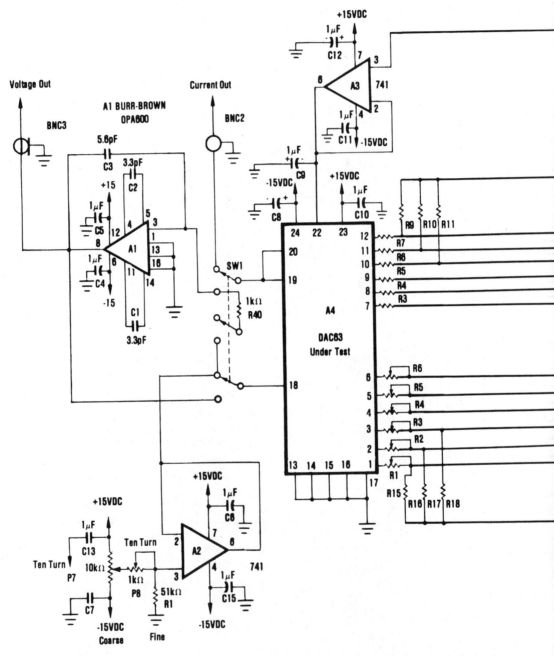

Circuit using the DAC63 to generate a staircase waveform which can be examined to evaluate the glitch performance.

Fig. 4-7

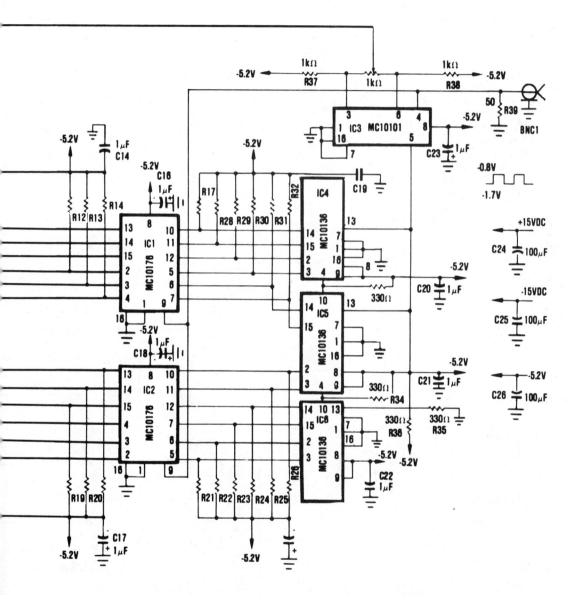

Glitch Measurement

DAC63

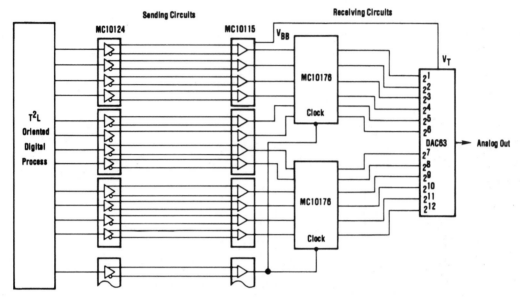

High speed DAC receiving data from a remote location.

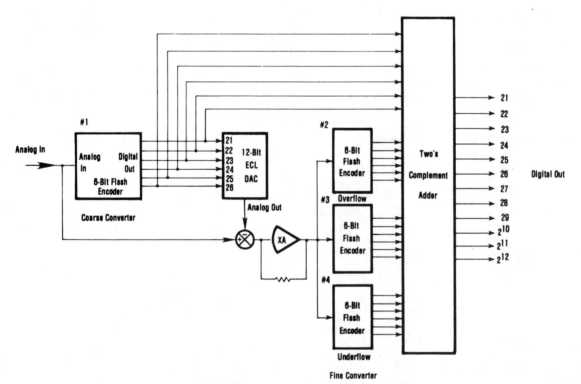

Fig. 4-8   Block diagram of 250 nsec analog-to-digital converter.   (BB)

Testing DA Converters

Testing digital-to-analog converters

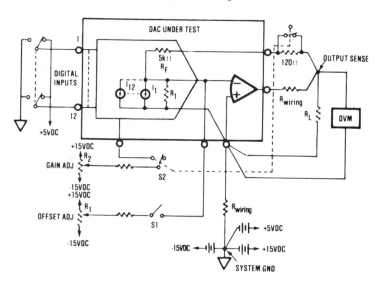

Fig. 4-9   Static DAC test circuit.   (BB)

Testing DA Converters   Testing DA Converters

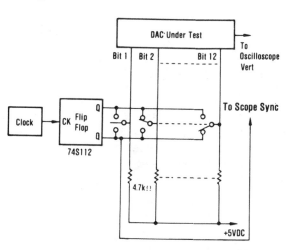

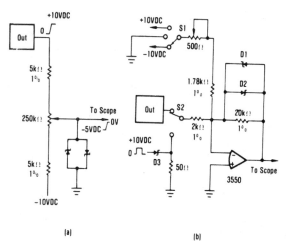

Simple low-cost DAC tester uses a flip-flop to generate complementary logic drive. SPDT switches with a center-off position and a pull-up (or pull-down) resistor are used to select code carries or full scale changes.

(a) A resistive divider to an offsetting voltage is used to produce a null. This circuit has a gain of 1/2.

(b) X10 gain amplifier with a method to verify the settling time of the gain amplifier.

Low cost DAC tester.   Circuits for measuring settling time.

Fig. 4-10   (BB)   Fig. 4-11   (BB)

Testing digital-to-analog converters

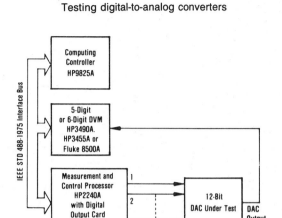

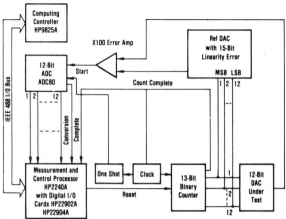

Automatic DAC test circuit
using a computing controller and a DVM.

Replacing the DVM of Figure 3 with this high-speed digitizing circuit and using the high-speed interrupt capability of the computing controller reduces the measurement period to less than 100μsec.

Fig. 4-12   High speed digitizing circuit.   (BB)

Testing DA Converters

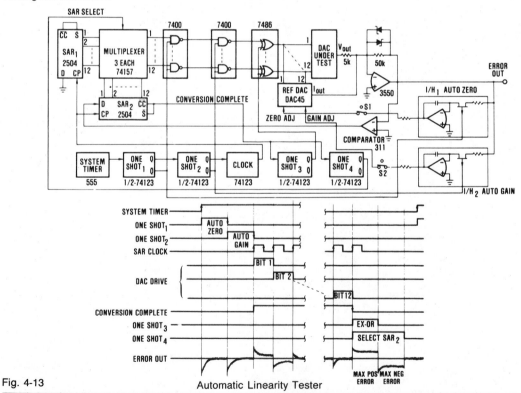

Fig. 4-13   Automatic Linearity Tester   (BB)

Testing DA Converters

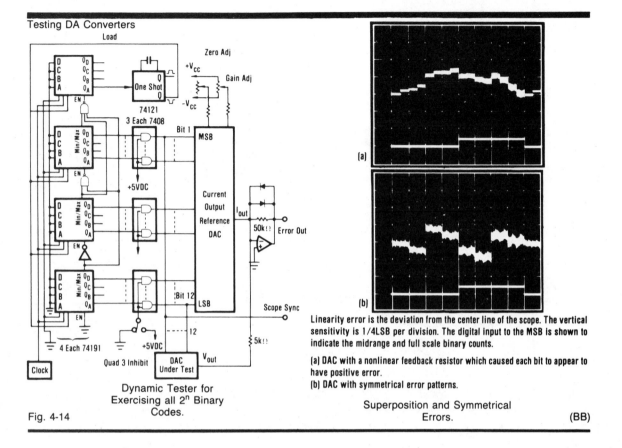

(a)

(b)

Linearity error is the deviation from the center line of the scope. The vertical sensitivity is 1/4LSB per division. The digital input to the MSB is shown to indicate the midrange and full scale binary counts.

(a) DAC with a nonlinear feedback resistor which caused each bit to appear to have positive error.
(b) DAC with symmetrical error patterns.

Superposition and Symmetrical Errors.   (BB)

Dynamic Tester for Exercising all 2^n Binary Codes.

Fig. 4-14

VFC32   Voltage-to-frequency converter interfacing circuits

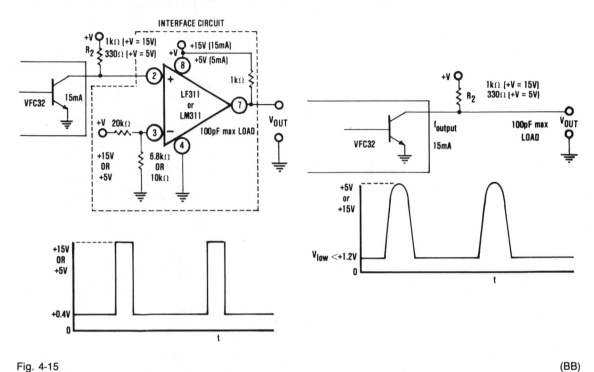

Fig. 4-15   (BB)

VFC32   Voltage-to-frequency converter interfacing circuits

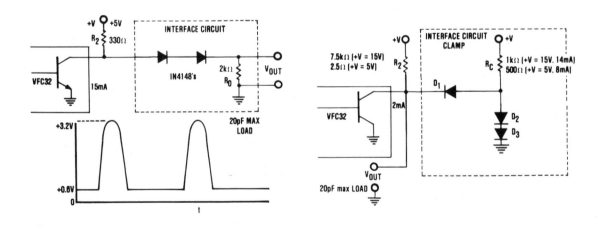

For additional VFC32 circuits see Chapter 1.

Fig. 4-16   (BB)

LM111

3584   High speed, high voltage DAC.

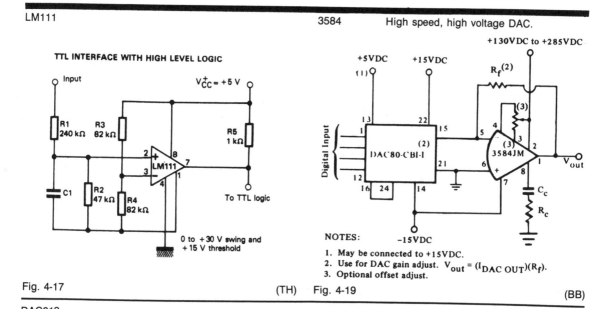

TTL INTERFACE WITH HIGH LEVEL LOGIC

NOTES:

1. May be connected to +15VDC.
2. Use for DAC gain adjust. $V_{out} = (I_{DAC\ OUT})(R_f)$.
3. Optional offset adjust.

Fig. 4-17   (TH)   Fig. 4-19   (BB)

DAC812

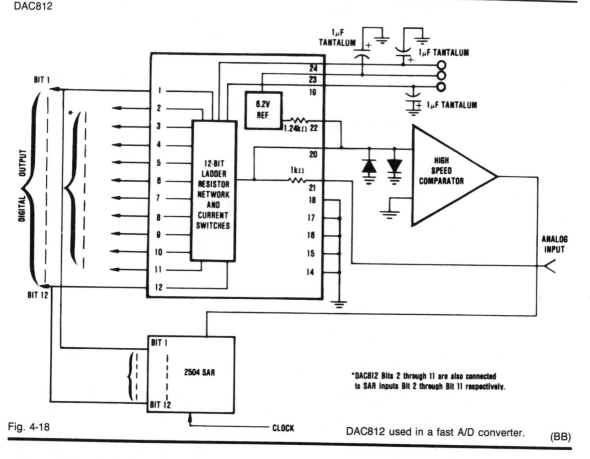

*DAC812 Bits 2 through 11 are also connected to SAR inputs Bit 2 through Bit 11 respectively.

Fig. 4-18   DAC812 used in a fast A/D converter.   (BB)

DAC63

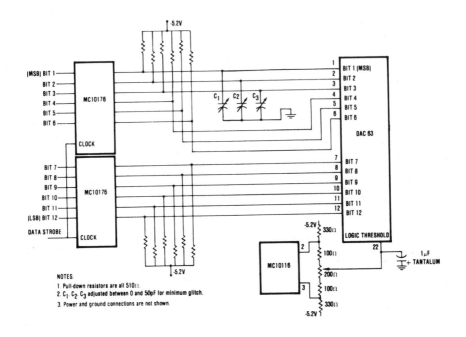

Fig. 4-20   DAC63 interface to input latches including glitch-adjust circuitry.   (BB)

DAC63

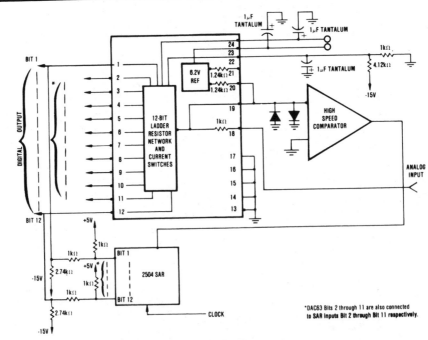

Fig. 4-21   DAC63 used in a fast A/D converter.   (BB)

GAMES PEOPLE PLAY WITH INTERSIL'S A/D CONVERTERS

This application note is a collection of application circuits, mostly small or special purpose circuits that could not justify publication in this form alone, but together constitute a useful compendium. The circuits have been loosely categorized into four groups for convenience in reference, as follows:

☐ Input Games: Circuits which alter the input signal(s) in some (preferably) unusual or non-obvious way, to achieve an unexpected or non-standard function.

☐ Converter Games: Circuits involving some alteration to the operation of the converter, to achieve some result other than the normal direct linear voltage conversion.

☐ Output Games: Games played with the output of the converter, to alter or add to the display, or add a computer interface or alarm, for example.

☐ Mixed Games: Games that cannot easily be categorized into one of the other types, or that involve more than one of the above activities.

Since many of the circuits are based on only a few of the wide range of available Intersil converters, a Schedule (Table) is provided to show to which converter circuits the applications relate, either directly or with modifications. It is assumed throughout that the "normal" operation of the devices is known to and understood by the reader. If this is not the case, the perusal of the relevant

Table 4-1. Schedule of Games.

NO.	TITLE	APPLICABLE DEVICES (ICL)										CATEGORY OF GAME
		7104	7106	7107	7109	7116	7117	7126	7135	7136	7137	
1	LCD Annunciators		Y			Y		Y	Y			Output
2	Decimal Point Drive		Y			Y		Y	Y			Output
3	Low Battery Detector		Y	Y		Y	Y	Y	Y	Y		Output
4	Blank Display on Low Battery		Y	Y		Y	Y	Y	Y	Y		Output
5	LED Brightness Control		Y			Y			Y			Output
6	Leading Zero Blanking		Y	Y		Y	Y	Y	m	Y	Y	Output
7	Instant Continuity on Ohms		Y	m	m	m	m	Y	m	Y	Y	Output
8	High Voltage Display Driving		Y	Y		Y	Y	Y		Y	m	Output
9	DVM Circuits		Y	Y		Y	Y	Y	Y	Y	Y	Input
10	Battery Operated Tachometer	Y	Y	Y	Y	Y	Y	Y	Y	Y	Y	Input
11	Examining your Navel	m	Y	Y	Y	m	m	Y	m	Y	Y	Input
12	Measuring h_{fe}		Y	Y	Y			Y	m	Y	Y	Input
13	Running off 1.5V Battery		Y	m	m	Y	m	Y	m	Y	m	Mixed
14	Conversion Status Signal		Y	Y				Y	m	Y	Y	Output
15	Switching Preamp		Y			Y		Y		Y		Input
16	Capacitance Meter		Y	m	m	Y	m	Y	m	Y	m	Input
17	Eliminating Overrange Hangover	Y										Converter
18	Auto-Tare Circuit	m										Converter
19	3¾-Digit Meter (LCD/LED/VF)	m			Y							Converter
20	¼ Count Resolution Meter	m			m				Y			Converter
21	A Logarithmic A/D Converter		Y	Y	Y	Y	Y	Y	Y	Y	Y	Converter
22	No-Ref-Cap Circuits	m	Y	Y	Y	Y	Y	Y	Y	Y	Y	Converter
23	Low Noise Preamp Circuit		Y	Y	Y	m	m	Y				Converter
24	Current I/P Converter Circuit		Y	Y	Y	m	m	Y				Converter
25	BCD Output from 7-Segment O/P		Y	Y		Y	Y	Y		Y	Y	Output

Note: "Y" means that the circuit is shown for the device listed (although frequently some component values may need alteration), whereas "m" signifies that some modification is required to use that device. The "ICL7104" is to be considered a chip pair, with the required ICL8052 or ICL8068 not listed, but assumed.

A047

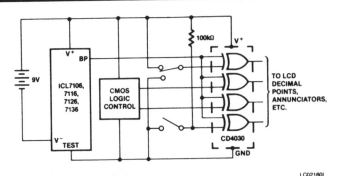

LC021801

Logic-Controlled LCD Annunciators. These circuits can be used to show any desired messages on the same display as the digital output

Fig. 4-22

(IN)

A047

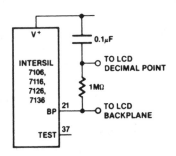

LC021701

DC-less Decimal Point Drive. Although the circuit of Figure 1 can provide fixed annunciator drive, it is relatively expensive. A crude drive method (shown in A023) is to tie the LCD segment to COMMON, but this applies a significant DC voltage to the display which can lead to problems. In any case it cannot be used if COMMON is grounded externally, etc. This circuit, suggested by a French correspondent, avoids those problems, and acts as a rough "low-battery" detector also. Note that by the time the segment disappears, the reference voltage on COMMON has altered significantly, and that the detection point is a function of the individual display

(IN)

Fig. 4-23

data sheet(s) and of the other application notes listed at the end of this one is recommended.

Evaluation Kits are available for the ICL7106, 7107, 7126, 7135, and 7136, containing the IC, a PC board, the required basic passive components, and in some cases a display, in others various ancillary ICs, together with the required instructions, etc. Most of the simpler circuits shown here can be easily built on the appropriate EV/Kit with a little component juggling and occasional trace cutting. Several of the kits have "breadboarding" areas to facilitate the addition of extra components.

A047

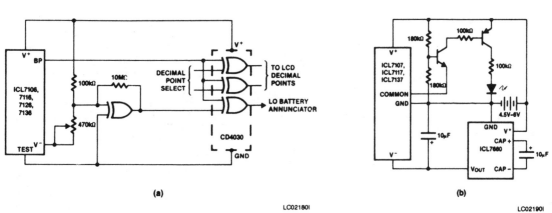

(a)

LC021801

(b)

LC021901

Low battery detector. The threshold is set by the resistor ratio against the internal logic regulator in each case. The remaining gates can be used for annunciator drivers, etc. as shown

Fig. 4-24

(IN)

A047

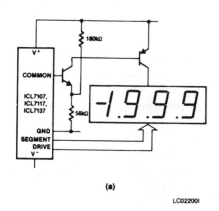

(a)

LC022001

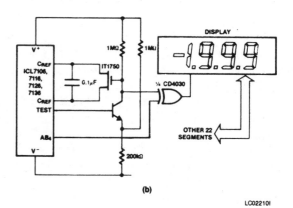

(b)

LC022101

Blank display on low battery. Sometimes no reading is better than a (possibly) bad one. The same type of circuit can be used to cover other conditions also. The LCD circuit forces an "overrange" reading by shorting the reference input, and inverts the "1", leaving only a possible polarity indication.

Fig. 4-25

(IN)

The circuits shown here come from various and sundry sources. Many came directly from users of the devices, and many others from questions or suggestions from them, still others came from our engineering and applications people around the world.

Enough of this! Without more ado, LET THE GAMES BEGIN!

A047

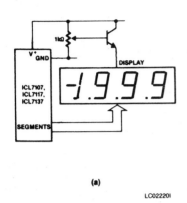

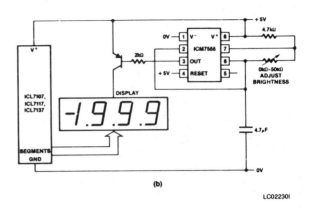

(a)   (b)

LC022201   LC022301

Controlling LED brightness. The circuit in (a) applies a variable voltage to the anodes of the LEDs, and can lead to mismatch problems, while that of (b) uses a timer circuit to control the duty cycle, needing good supply bypassing. A counter driven by the oscillator could be used instead of a timer

Fig. 4-26   (IN)

High Voltage Display Driving

A vacuum-fluorescent display panel can be activated by three D1805A high voltage drivers which are directly driven by an ICL71X7 3½-digit A/D converter. The ICL71X7 normally drives a common-anode LED display and its outputs pull-down to turn on a segment. The D1805A acts as an inverter in this application to illuminate display segments when its inputs are pulled down. Note the grid and filament connections to the display panel in Fig. 4-29.

A gas-discharge type display is shown in Fig. 4-30 being activated by three DI220 display driver circuits which are driven by an ICL71X6 A/D converter. The ICL71X6 is designed to drive an LCD display. In this application, the BackPlane signal is buffered by a pnp emitter-follower and, when low, enables the high voltage drivers by providing a ground return for them.

A high level on the input to the DI220 when BackPlane is low will turn on a display segment. Thus, only when a segment line is out of phase with BackPlane will the segment be on. In this mode, the display will be operating at 50% duty cycle.

Battery Operated Tachometer

The problem of getting a reading in rpm from an input frequency is solved by the use of a frequency-to-voltage converter at the front end of an A/D converter.

The frequency-to-voltage conversion is accomplished by a CMOS timer which generates a constant pulse width waveform at its output and a micropower CMOS op amp which integrates the timer pulses. Operating the timer from the internal reference voltage of the A/D converter eliminates the need for a second reference because of the rail-to-rail output swing of the timer.

A047

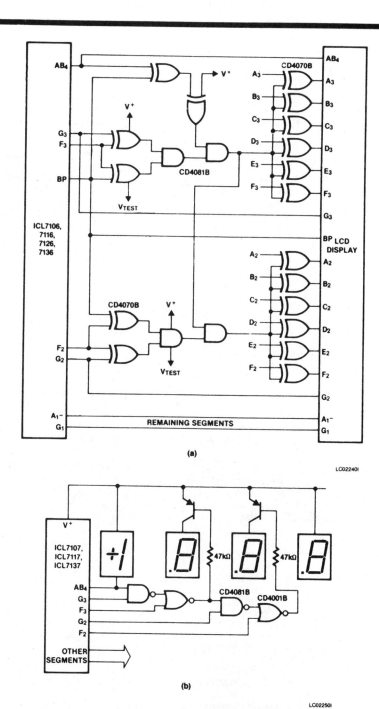

(a)

LC022401

(b)

LC022501

Leading zero blanking for the ICL71X6 and 71X7. The circuits detect the "segment f and not-segment g" condition, and blank the display accordingly

Fig. 4-27

(IN)

A047

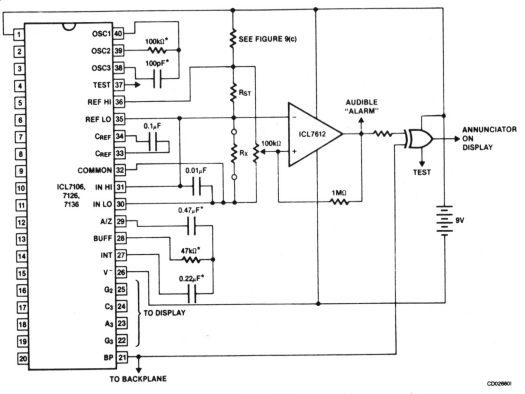

*ICL7106 only. See data sheets or A052 for correct values of these components.

Instant continuity when measuring resistance. The output can be used to drive an audible alarm, etc. if desired. The continuity annunciator or audible signal will indicate regardless of normal measurement cycle if input is less than set proportion of full-scale input. For use with ICL7107 or ICL7109, different annunciator connection will be needed, of course. To use with ICL7116, 7117, or 7135 reverse positions of known R_{ST} and unknown R_X, swap IN HI and REF HI, IN LO and REFLO, and invert sense of output. The ICL7135 will also need a supply to drive the resistors

Fig. 4-28   (IN)

The input to the A/D converter is given by:

$$V_{IN} = \left(\frac{rpm}{60}\right)(t_{pw})\,(V_R)\,(E)\left(\frac{R4}{R3}\right)$$

Where t_{pw} = timer pulse width = 1.1 R2C2
V_R = ICL7106 Reference Voltage = 2.8 V
E = Number of events per revolution = 2 for this example

The reading of the A/D converter is given by:

$$n = \left(\frac{V_{IN}}{V_R}\right)\left(\frac{R8 + R7}{R8A}\right)$$

$$n = \left(\frac{rpm}{60}\right)(1.1\ R2C2)\ (E)\left(\frac{R4}{R3}\right)\left(\frac{R8\ +\ R7}{R8A}\right)$$

The factor E relates to the number of pulses per revolution from a magnetic or optical sensor, the number of blades on a propeller, or the number of point closures per revolution in an automotive application. R6 is needed only if the A/D converter is adjusted for 200 mV full-scale.

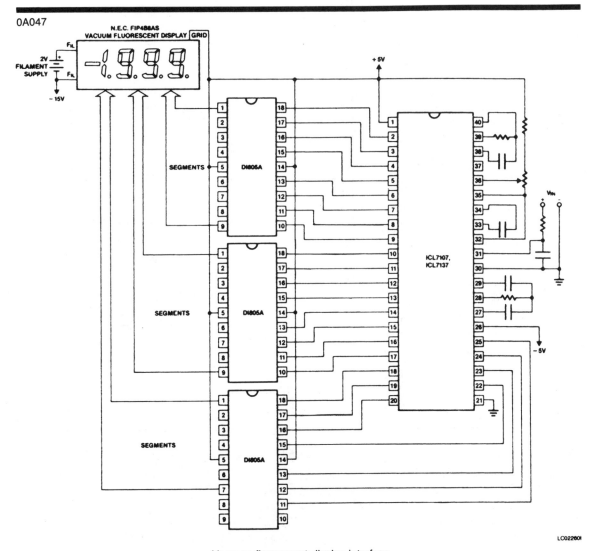

Vacuum flourescent display interface

Fig. 4-29   (IN)

A047

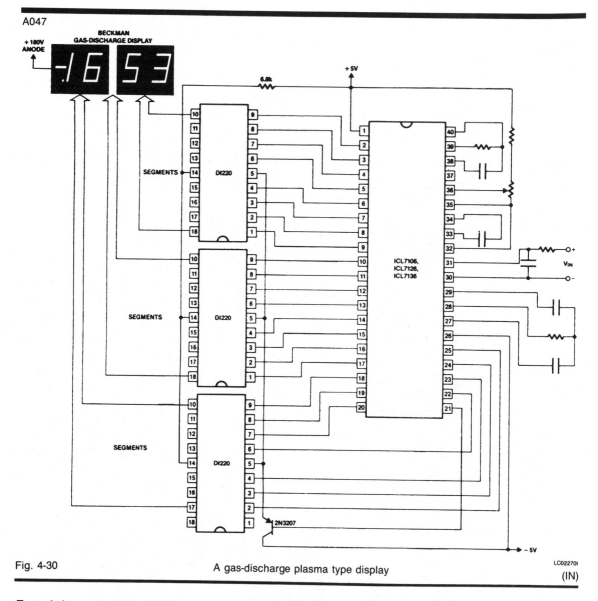

Fig. 4-30   A gas-discharge plasma type display   LC022701
                                                                                    (IN)

Examining your Navel (alias Looking at the Beam in your Own Eye . . .)

One common application question concerns using an ICL7106/7 family device to monitor its own supplies. Obviously this won't work at all if the supply gets too low, but above this level the circuits of Fig. 4-33 will do just that. The ICL7107 circuits can also be used for the ICL7109 and ICL7135.

Measuring h_{fe}

The circuit of Fig. 4-34 sets the emitter current of the DUT via R3, and measures the ratio of the resultant base and collector currents, monitored in R1

A047

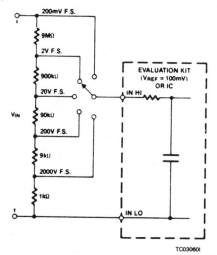

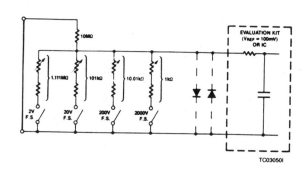

*CAUTION: High voltages can be lethal. Proper operating precautions must be observed by the user. Intersil assumes no liability for unsafe operation.

(a) Multirange Voltmeter

*200mV F.S. for all switches open

(b) Multirange Voltmeter, Alternative Arrangement

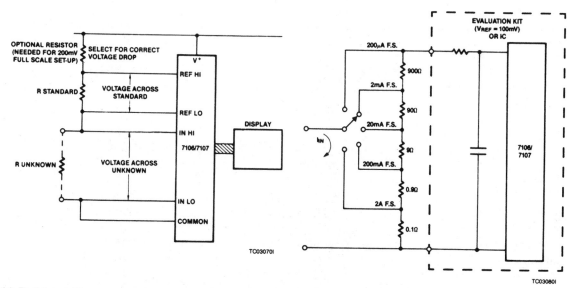

(c) Resistance Measurement. The optional resistor can be replaced by a diode string

(d) Multirange Current Meter

DVM circuits. Voltage and current measurements for metering.
For auto-ranging circuits see A046 for the 3½-digit devices, and A028 for the 4½-digit parts

Fig. 4-31

A047

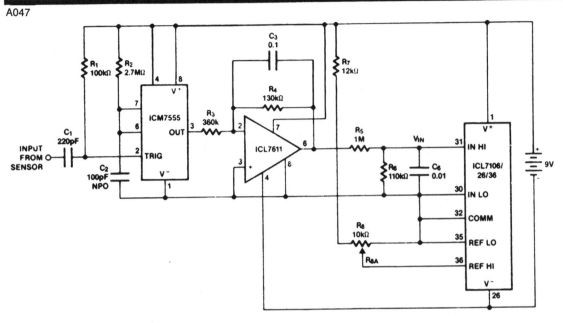

Tachometer using ICL7106 family A/D

Fig. 4-32

(IN)

and R2 at the REFerence and INput terminals respectively. A fixed voltage and R3 set the emitter current, hence the slightly smaller collector current and the input voltage, while the reference voltage varies as

$$\sim \frac{1}{h_{fe}}.$$

Although high betas cause overrange readings, the circuit is not actually overloaded, while with low betas the high reference gives a short de-integrate cycle, but still accurate readings.

For the resistor values shown, full-scale is 199.9; increasing R1 by a factor of 10 will give 1999 full-scale. R1 and R2 should be 0.1% or better, but R3 is not critical. Although Fig. 4-34 shows separate npn and pnp circuits, they could be combined with a range-switching arrangement. Minor component value changes will accommodate the ICL7126/7136.

Running off 1.5 V Battery (with aid of Voltage Converters)

Two ICL7660 voltage converters quadruple the 1.5 V single cell battery voltage into 6 V, sufficient to power the ICL7126/7136 A/D converter and an ICL8069 voltage reference. The latter is needed because the internal reference needs over 6.5 V to operate correctly. The battery current is typically less than ¾ mA, and batteries up to 3.5 V may be used. CR1 may be required for batteries over

3.0 V. The ICL7106 and ICL7116 can use the same circuit, but with higher battery drain. A similar arrangement can also be used with the ICL7109 and ICL7135, and also the ICL7107/7117 although LED display currents are usually too high.

Conversion Status Signal

Many applications of the ICL7106/7 family would be facilitated by a signal similar to the BUSY or STATUS signal on such converters as the ICL7109 and 7135. The simple circuit shown here will accurately signal the end of conversion, for use in multiplexing, auto-ranging, etc. Note that it will only work with the reference input near V⁺ and with reference voltages of less than a few hundred mV. Also the reference is disturbed by the current drawn, so ratiometric use is not recommended, and the beginning of conversion signal will be appreciably delayed. The good news is that the roll-over error discussed in A032 for reference not at COMMON does not apply to this circuit. See A046 for a variant of this.

A047

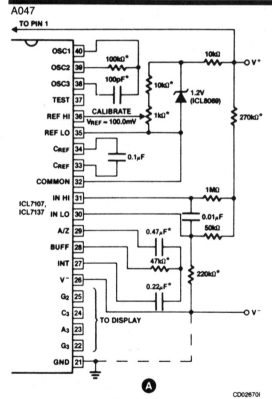

CD026701

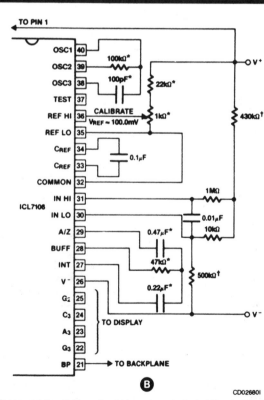

CD026801

*ICL7106/7 only. See data sheet for values for other parts.

†These resistor dividers should be set up so that at "end-of-life" for the supply, IN LO is about 2.8V below V⁺, and correct division occurs to IN HI-IN LO.

(a): Voltage Monitor, LED Version using 7107. Dotted connection should be used for supplies of 4 V-6 V. For higher supplies, provide regulated 5 V between V⁺ and GND.

(b): Voltage Monitor, LCD Version, for Voltages over 6.5 V (to 15 V max). Circuit uses internal reference. For <6 V, use ICL8069 reference.

Fig. 4-33A & B

(IN)

A047

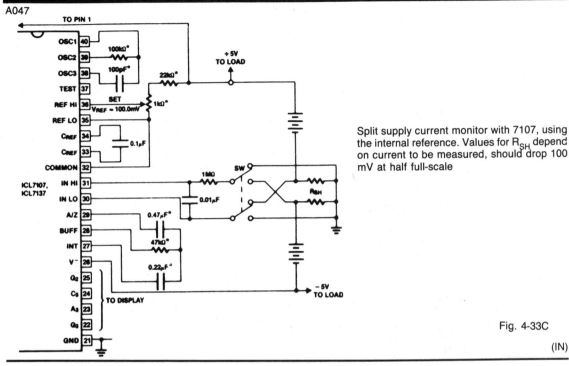

Split supply current monitor with 7107, using the internal reference. Values for R_{SH} depend on current to be measured, should drop 100 mV at half full-scale

Fig. 4-33C

(IN)

A047

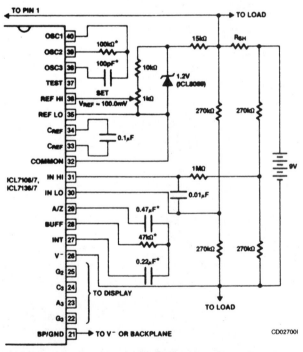

Single supply current monitor using external reference. Circuit can use ICL7106 or 7107. Resistor dividers must match very accurately to ensure accuracy. R_{SH} depends on current to be monitored, division ratio of dividers

Fig. 4-33D

*ICL7106/7 only. See data sheet for values for other parts.

(IN)

A047

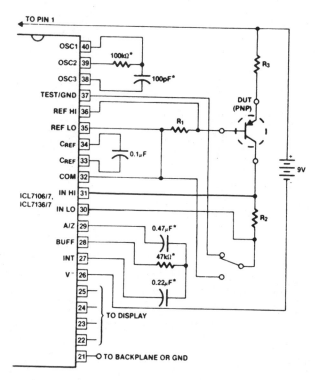

$I_E(I_C)$	$1\mu A$	$10\mu A$	$100\mu A$	$1mA$
R_1	20M	2M	200k	20k
R_2	200k	20k	2k	200Ω
R_3	2M	200k	20k	2k

Digital h_{fe} tester. The switch shown in the pnp version (a) permits testing at $V_{CD} \sim 0.5$ V or ~ 3.5 V. This cannot be done on the npn tester (b), however, without exceeding the common-mode range of the input

(a)

CD027101

Fig. 4-34

(IN)

A047

LCD DISPLAY

Using ICL7660 to run ICL7126/36 from 1.5 V battery

LC022901

Fig. 4-35

(IN)

A047

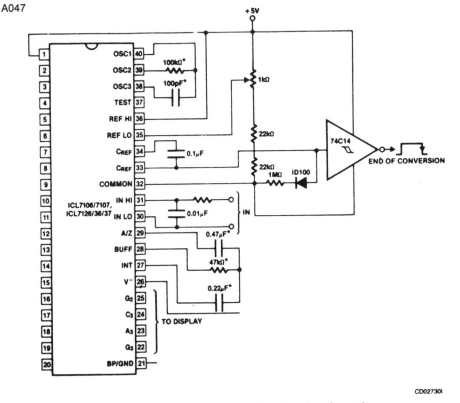

*ICL7106/7 only. See data sheet for values for other parts.

CD027301

Fig. 4-36   Simple end-of-conversion detector   (IN)

Switching Preamp

The low-signal resolution of even the best A/D converters is limited by the effective noise voltage at the inputs, generally dominated for the ICL7106/7 family by the noise voltage trapped on the auto-zero capacitor.

Low cost preamplifiers usually contribute wildly excessive offset errors, but the circuit here uses the backplane drive output of an ICL7106 (or 7116, 7126, or 7136) to synchronously switch a differential amplifier through a pair of polarity reversing analog switches. The input always sees the signal in the same polarity and magnitude, but the preamp offset is inverted with a 50% duty cycle at the A/D input so that it averages zero over the integrate cycle. The switching is performed at a 60 Hz rate, and performs excellently down to 20 mV full-scale (10 μV/count).

Most low cost dual amplifiers are suitable, but it is important that the negative and positive slew rates be reasonably close. Op amps with crossover distortion (such as the LM124/324) should not be used. See Fig. 4-37 and also application notes for alternative preamplifier circuits.

Capacitance Meter Based on 3½-Digit A/D

The circuit charges and discharges a capacitor at a crystal-controlled rate, and stores on a sample-and-difference amplifier the change in voltage achieved. The current that flows during the discharge cycle is averaged, and ratiometrically measured in the A/D using the voltage change as a reference.

Range switching is done by changing the cycle rate and the current metering resistor. The cycle rate is synchronized with the conversion rate of the A/D by using the internal OSCillator (externally divided) and the (internally divided) BackPlane signals. For convenience in timing, the switching cycle takes 5 counter states, although only four switch configurations are used. Capacitances up to 200 µF can be measured, and the resolution on the lowest range is down to 0.1 pF.

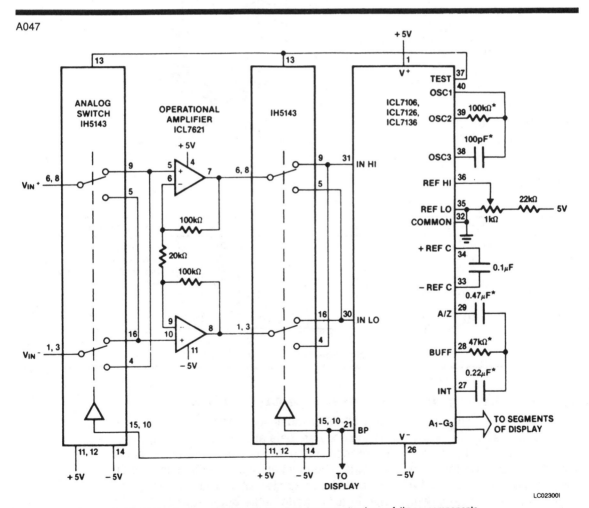

A047

*ICL7106/7 only. See data sheets or A052 for correct values of these components.

LC02300I

Fig. 4-37   Switching preamp with gain of 10   (IN)

A047

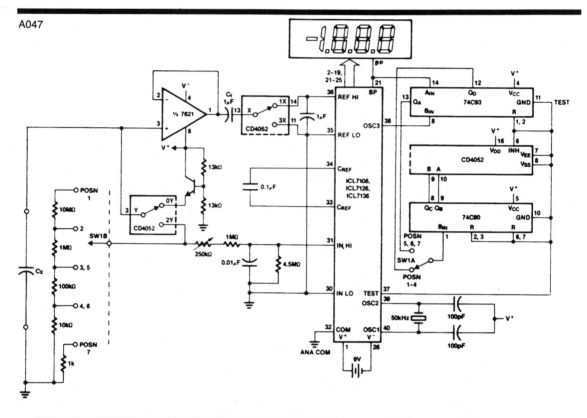

	SWITCH 1A	SWITCH 1B	MAX C		COUNTER/SWITCH PHASES	
1	10MΩ	6kHz	200pF	0Y	00	Charge C_X
2	1MΩ	6kHz	2nF	1Y	01	ΔV_{CX} on C_{REF}
3	100kΩ	6kHz	20nF	2Y	10	Discharge C_X thru R_{net}
4	10kΩ	6kHz	0.2µF	3X	11	Reset C_t to zero
5	100kΩ	60Hz	2µF			
6	10kΩ	60Hz	20µF			
7	1kΩ	60Hz	200µF			

Fig. 4-38   Capacitance meter built on 3½-digit meter   (IN)

Eliminating Overrange Hangover of Dual-Slope A/D Converters

Intersil's two chip 4½-digit and 14-16-bit dual-slope A/D converters have eliminated most of the hassle of building accurate data acquisition systems, dvms, dpms, etc. However, the published standard hook-up circuits can, under some circumstances, give inaccurate readings. The most annoying of these is after an overload, when the residual voltage on the integrator is transferred, as a transient, to the auto-zero system, which then in turn has to settle back down.

In any system with a multiplexed input, having succeeding channels disturbed by an overload condition on one channel can be a severe problem.

This "hangover" can be eliminated by adding a simple 2 IC circuit to zero the integrator output at the beginning of the auto-zero cycle, as shown in Fig. 4-39A for the ICL8052/71 (C)03 device, and Fig. 4-39B for the ICL8052/7104-16 device. The circuit works by connecting the comparator output to an inverting input on the buffer during the first portion of the auto-zero time, and then (as normally) to the auto-zero capacitor for the remainder (Fig. 4-39C).

A047

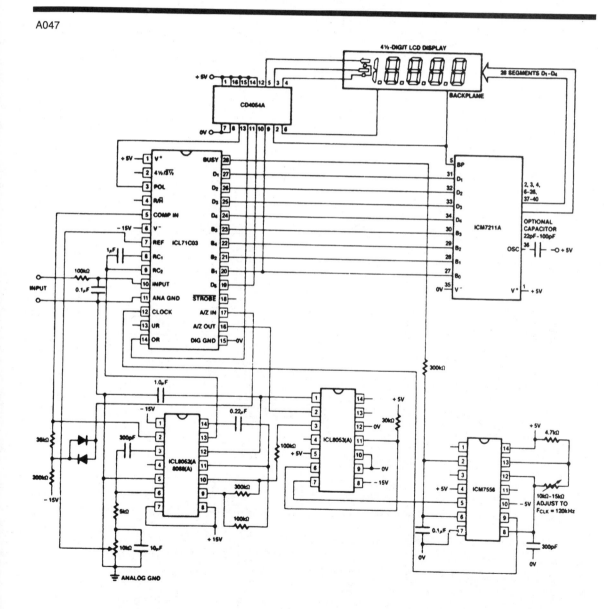

Fig. 4-39A   4½-digit LCD dmp with zero integrator phase   (IN)

A047

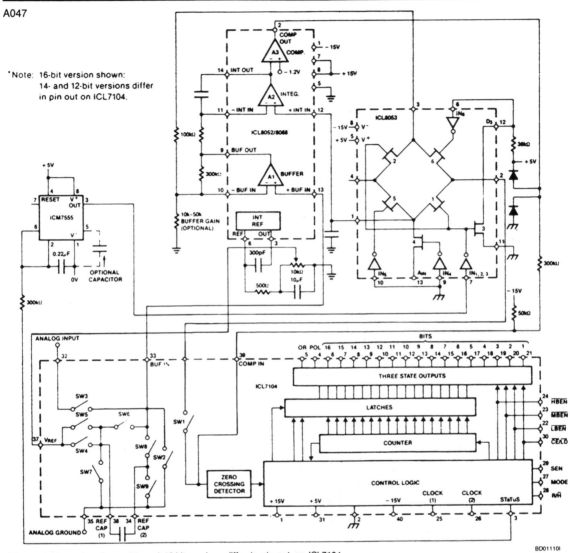

*Note: 16-bit version shown:
14- and 12-bit versions differ
in pin out on ICL7104.

Note: 16-bit version shown: 14- and 12-bit versions differ in pin out on ICL7104.

BD01110l

8052A (8068A)/7104 16-/14-/12-Bit A/D Converter Functional Block Diagram with Gross Overrange Protection Zero Integrator Phase Circuit

(IN)

Half of the ICM7556 controls the timing, the other half forming the clock oscillator, and the ICL8053 performs the switching. This circuit is used, rather than other more normal switch devices, because of its low charge injection into the auto-zero capacitor.

The Zero Integrator time can be set initially at ⅓ to ½ the minimum auto-zero time, but if an "optimum" adjustment is required, look at the comparator output with a scope under worst-case overload conditions. The output of the delay timer should stay low until after the comparator has come off the rail, and is in the linear region (usually fairly noisy).

Under non-overload conditions, a small "zero integrator" phase is desirable, but the full time given here is not necessary. A p-channel MOSFET across the timing resistor can be used to reduce the time constant if overrange has not occurred. Alternatively, a 3-diode network can be used, or a 2-input OR gate, as in Fig. 4-39D.

The delay circuit also introduces a delay from the beginning of the conversion interval before the Zero Integrator phase is enabled; this delay should not exceed the input integrate time. If circuit modifications are made, this should be checked.

A047

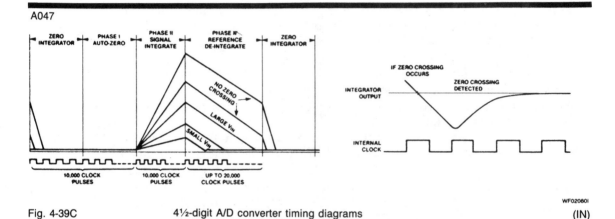

Fig. 4-39C   4½-digit A/D converter timing diagrams   (IN)

A047

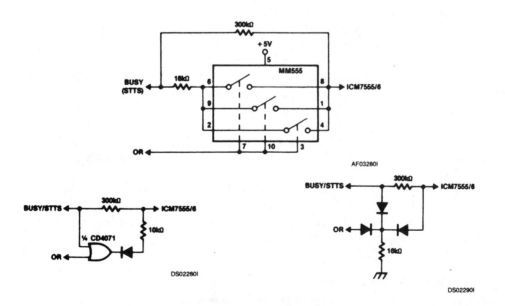

Fig. 4-39D   Timing gross overvoltages   (IN)

A/D Converter and Counter make Auto-Tare Weighing Electronics

Using Intersil's ICL7109 12-bit A/D and ICM7217 4 decade display driving counter, this circuit provides simple pushbutton tare for a weighing machine and requires only a single 5 V supply. With any tare weight on the scale, the press of a button zeroes the display, which then reads weight added or removed.

Operated in its handshake mode, the ICL7109 dual-slope converter gives a HBEN pulse at pin 19 one clock period after detection of zero crossing. By the nature of dual-slope conversion the zero crossing, hence the HBEN pulse, will move N clock periods later as the input is increased N bit values, and vice versa (see Fig. 4-40B). Each conversion cycle lasts 8192 clock periods.

The 7109 input can be derived from a bridge connected load cell powered from the same 5 V supply as the ICL7109 and its reference, so the ratiometric operation of the 7109 rejects supply variations.

The ICM7217 4 decade bidirectional counter has display latches enabled by STORE (pin 19) which drive a multiplexed LED display. The ZERO output (pin 2) indicates the zero state of the counter, which also has a RESET input (pin 14). A separate compare register is preloaded via diodes from the display digit selects to the BCD port when Load Register (pin 11) is high. EQUAL (pin 3) goes low when register and counter are equal.

At power-up, the 10 μF capacitor on Load Register (pin 11) preloads the compare register to 4096. The counter UP/DOWN line is controlled by a flip-flop (½ 74LS74) alternately set and reset by ZERO and EQUAL so the counter cycles up and down between zero and 4096, taking 8192 clock periods per cycle.

With a steady weight on the scale, the HBEN pulse from the 7109 activates the STORE input of the 7217 every 8192 clock periods to give a steady, but initially random display. When the TARE switch is closed, the counter is reset by HBEN, zeroing the display. Even when the TARE switch is opened, the display remains zero because, with an unchanged weight, HBEN enables the display latches each time the counter passes zero (solid line in Fig. 4-40B).

If weight corresponding to N bit values is added to the scale, HBEN moves N counts later, hence N will be latched and displayed (dotted line in Fig. 4-40B). Conversely, if N bit values of weight are removed, $-$ N is displayed. The minus sign is obtained by latching the state of the UP/DOWN line at HBEN in the second flip-flop.

The only restriction is that the converter input must not change polarity. If necessary, the POL output (pin 3) of the 7109 can be latched to warn of such an erroneous condition.

For more than 4096 count resolution, the 7109 can be replaced by the ICL8052A/ICL7104 2-chip converter set with up to 16-bit resolution. The ICM7217 counter is cascadable. The system can clearly also operate with force balance type weighing systems or other types of input signal.

3¾-Digit (± 4095 Count) Meter with LED, LCD or Vf Display

Using an ICL7109 12-bit A/D, an ICM7224, 7225 or 7236 counter/display driver and two 4000 series CMOS chips you can build a 3¾-digit (4095 count) auto polarity

A047

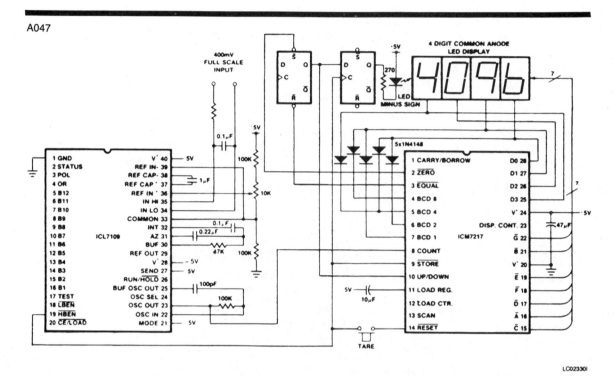

Auto-tare weighing system for A/D converter

Fig. 4-40A   (IN)

A047

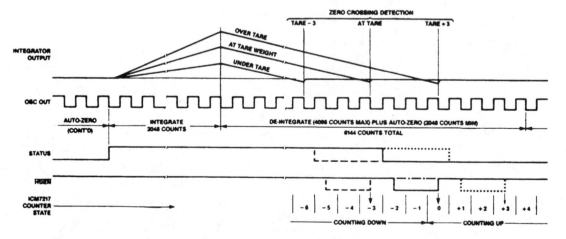

Auto-tare weighing system waveforms

Fig. 4-40B   (IN)

digital meter. The choice of counter determines the choice of display. ICM7224 drives liquid crystal displays, ICM7225 driven common anode LEDs and ICM7236 drives bright, green vacuum fluorescents.

Figure 4-41A shows the system schematic. The ICL7109 is run in its handshake mode (MODE wired high). The analog input is completely conventional and the analog connections are not shown in detail. Waveforms are shown in Fig. 4-41B.

The ICL7109 takes 8192 clock cycles to complete a conversion, 2048 for auto-zero, 2048 for signal integrate and 4096 for reference de-integrate. STATUS goes high at the beginning of signal integrate and comes low 1½ cycles after the zero crossing, giving a total of 2050 + N pulses of the clock while status is high, where N is the digital reading of the A/D. Therefore, by counting clock pulses while STATUS is high and subtracting 2050, the reading can be obtained.

The ICM7224/7225/7236 display driving counters have identical input configurations. They contain a 4½ decade counter with CLOCK, RESET and COUNT INHIBIT inputs and separate display latches activated by STORE. These latches in turn drive, via decoders, non-multiplexed 7-segment displays.

When STATUS goes low, RESET of the CD4040 is taken high and the bi-stable composed of gates 3 and 4 is set, taking COUNT INHIBIT low. At the beginning of integrate STATUS goes high, removing RESET from the CD4040 and allowing it to start counting. Gate 3 detects the count of 2050 and resets the bi-stable, removing COUNT INHIBIT and allowing the counter to count clock pulses. When STATUS goes low after zero crossing the bi-stable is again set, and further

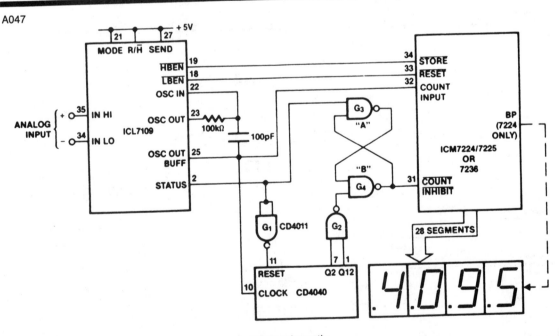

A047

Fig. 4-41A   Block schematic   (IN

A047

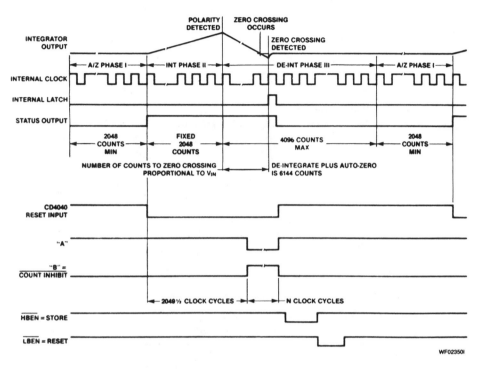

Fig. 4-41B   Operating waveforms   (IN)

counting inhibited. The counter has now counted any pulses over 2050, hence now contains the required reading. This is transferred to the display latches by $\overline{HBEN}$ from the 7109. The counter is reset by $\overline{LBEN}$ and the circuit is now ready for the next conversion cycle. The binary outputs of the 7109 are not used.

Of course, other counters can be used instead of the ICM7224/7225/7236, but these are extremely convenient in their ability to drive displays directly. The ICM7217 could be used with LED displays and its compare register could also be employed to provide an over or under limit indication.

Should greater resolution (up to 65536 counts) be required, the ICL7109 may be replaced by the ICL8052/8068 + ICL7104 14- or 16-bit converters, which have the same output interface as the 7109. The ICM7224/25/36 counters are cascadable. In this case the CD4040 counter chain would have to be extended to the appropriate length for 14- or 16-bit operation, as in these cases 8194 and 32770 counts respectively would have to be subtracted.

Because the display drive is non-multiplexed, it is convenient to mount the counter/display driver package on the same board as the displays. Only 4 signal connections (i.e., COUNT, $\overline{COUNT\ INHIBIT}$, $\overline{STORE}$ and $\overline{RESET}$) are then required from the main circuit board to the display board. Moreover, by interchanging display boards LED, LCD or Vacuum Fluorescent meters could be made using the same main circuit board.

5000 Count Weighing with ¼ Count Resolution

Low cost digital scales are readily built using a strain gauge transducer, and ICL7650 chopper stabilized CMOS amplifier, a dual-slope A/D converter such as Intersil's ICL7106/7 (3½-digit), ICL7109 (12-bit binary) or ICL7135 (4½-digit) single chip CMOS devices, plus, if necessary, a display driver. To be approved for trade, however, the electronics must resolve to ¼ of a displayed increment. This can be simply arranged by replacing the display driver with a display driving counter and a handful of standard SSI CMOS.

The ICL7135 4½-digit dual-slope A/D chip has 3 phases of operation: auto-zero, integrate and de-integrate. In the auto-zero phase, offsets are measured and nulled and the BUSY output is low. During integrate, the BUSY output goes high and the ICL7135 integrates the input signal for 10,000 clock cycles. During de-integrate, the reference is integrated until the integrator returns to its starting point (or zero crossing). If N is the digital reading, de-integrate lasts N + 1 clock cycles (the extra one is due to the fact that de-integrate actually ends on the next positive clock edge after zero crossing), and at the end of this, BUSY returns low. BUSY is therefore high for a total of 10,001 + N clock cycles, and clearly the reading can be determined by counting clock cycles while BUSY is high and subtracting 10,001.

In our case, we wish to display N/4, so that the 20,000 count conversion will yield a full-scale reading of 5000. This is done by delaying for 10,001 clock cycles after BUSY goes high, then enabling a counter clocked at ¼ the 7135 clock rate. When BUSY goes low this counter is halted, its contents transferred to the display, and reset ready for the next cycle. A small refinement is that the counter should increment at values of 2, 6, 10 etc., rather than at 4, 8, 12 etc. so that quantization error is symmetrical about the reading. A useful economy is to use the same counter for the initial 10,000 clock cycles and for the final counting. Suitable counters are Intersil's ICM7224 (driving LCD display), ICM7225 (driving LED display) or ICM7236 (driving vf display). Apart from the display driver, these are identical 4½-decade counters.

The circuit works like this; during auto-zero, when BUSY is low, the counter is disabled via its $\overline{\text{COUNT INHIBIT}}$ input. Flip-flop 3 is set, so that its Q output holds flip-flops 1 and 2 reset and gate 3 is enabled, passing 7135 clock pulses to the counter.

At the beginning of integrate, BUSY goes high and the counter begins to increment. Notice that the counter and flip-flops are incremented on the falling edge of clock, while the ICL7135 increments on the rising edge, avoiding race conditions. After 10,000 counts, the falling edge of $\overline{\text{CARRY}}$ clocks flip-flop 3 to its reset state. Flip-flops 1 and 2 are now enabled to divide the input clock by 4, while G3 blocks direct clock input to the counter, which has now rolled to zero.

The preset conditions of flip-flops 1 and 2 are such that the first rising edge of Q2 comes 3 clock cycles later, so the counter increments to 1 just before the ICL7135 increments to 2, and so on. At the end of de-integrate BUSY comes low, freezing the counter. The $\overline{\text{STROBE}}$ output of the ICL7135 consists of 5 pulses, one coincident with each digit select, during the first multiplexed output scan cycle after de-integrate ends. $\overline{\text{STROBE}}$ is used to set flip-flop 3, restoring initial conditions

A047

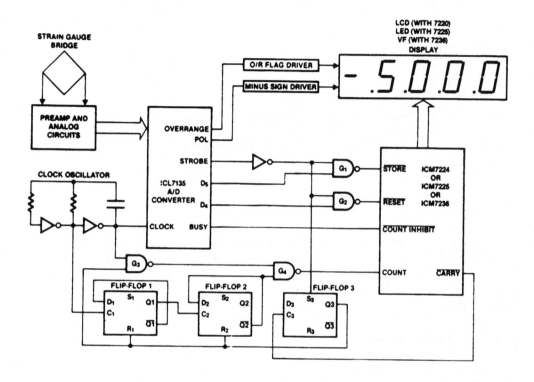

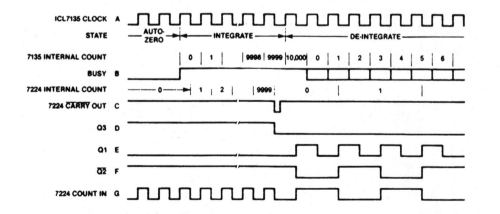

5000 count meter with internal ¼ count resolution

Fig. 4-42　　　　　　　　　　　　　　　　　　　　　　　　　　(IN)

ready for the next conversion. $\overline{\text{STROBE}}$ gated with D5 stores the counter contents in the display register, then $\overline{\text{STROBE}}$ with D4 resets the counter. Notice that the normal BCD outputs of the 7135 are unused.

A Logarithmic A/D Converter

An ordinary single chip dvm can be easily converted to display the logarithm of the ratio of 2 input voltages V_1 and V_2. The main restriction on operation is that $V_1 \geq V_2$.

Figure 4-43A shows the configuration. The modifications from the standard connection are the potential divider (ratio K) on the reference input, the signal input being the difference between V_1 and V_2, and the addition of resistor R_p in parallel with the integrator capacitor. The time constant of the integrator capacitor and R_p is given by:

$$\tau = C_{INT}R_p.$$

Referring to the waveforms of Fig. 4-43B, let us first calculate the final integrator voltage, V_{INT}. The aiming potential of the exponential is given by:

$$V_{ASSYMPTOTE} = \frac{R_p}{R_{INT}} (V_1 - V_2)$$

and the final integrator voltage is therefore:

$$V_{INT} = \frac{R_p}{R_{INT}} (V_1 - V_2) \left(1 - e^{\frac{-T}{\tau}}\right)$$

where T is the fixed integration period.

During de-integrate, the total swing of the exponential is given by:

$$V_{TOTAL} = V_{INT} + V_{REF} \frac{R_p}{R_{INT}}$$

and remembering that $V_{REF} = KV2$, the total exponential swing is:

$$V_{TOTAL} = \frac{R_p}{R_{INT}} (V_1 - V_2) \left(1 - e^{\frac{-T}{\tau}}\right) + \frac{R_p}{R_{INT}} KV2.$$

The integrator will actually cross zero when the exponential has reached

$$V_{FINAL} = V_{REF} \frac{R_p}{R_{INT}} = \frac{R_p}{R_{INT}} KV2.$$

Therefore, the time to zero crossing is given by:

$$T_{DE\text{-}INT} = \tau \ln \left(\frac{V_{TOTAL}}{V_{FINAL}} \right)$$

$$= \tau \ln \left(\frac{\dfrac{R_p}{R_{INT}} (V_1 - V_2)\left(1 - e^{\frac{-T}{\tau}}\right) + \dfrac{R_p}{R_{INT}} KV2}{\dfrac{R_p}{R_{INT}} KV2} \right)$$

$$= \tau \ln \left(\frac{(V_1 - V_2)\; 1 - e\left(\frac{-T}{\tau}\right) + KV2}{KV2} \right)$$

Now, if we are sneaky, and make $K = \left(\dfrac{1 - e^{\frac{-T}{\tau}}}{\tau}\right)$,

$$T_{DE\text{-}INT} = \tau \ln \frac{K(V_1 - V_2) + KV2}{KV2} = \tau \ln \frac{V_1}{V_2}$$

Now let us make $\tau = \dfrac{T}{2.3}$

where T is the integration period:

$$\text{Now } T_{DE\text{-}INT} = \frac{T}{2.3} \ln \frac{V_1}{V_2} = T \log_{10} \frac{V_1}{V_2}$$

so that the dvm will read 1,000 ($T_{DE\text{-}INT} = T$) when $V_1/V_2 = 10$, which is correct. The divider ratio, K, is:

$$K = \left(1 - e^{\frac{-T}{\tau}}\right) = 1 - e^{-2.3} = 1 - 0.1 = 0.9$$

which makes sense when you realize that the final integrator voltage during the integrate phase much reach 0.9 of the assymptote level.

Theoretically, the full-scale of the system is $V_1/V_2 = 100$ (i.e., when the log = 2) but noise will probably limit this to lower values. Note also that the accuracy of the system is no longer independent of passive component variations. The simplest set-up procedure is to ensure $K = 0.9$ (preferably using a pretrimmed divider) then, with $V_1 = 10\ V_2$, adjust R_p until the reading is 1,000. Examples of the use of logarithmic readings are photographic and chemical densitometry and colorimetry and audio decibel scales.

A047

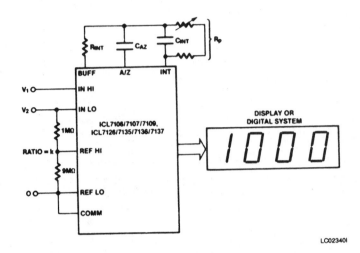

Circuit modifications for logarithmic operation. For ICL7116, ICL7117, and ICL7135, REF LO is already connected to ANALOG COMMON

Fig. 4-43A

(IN)

A047

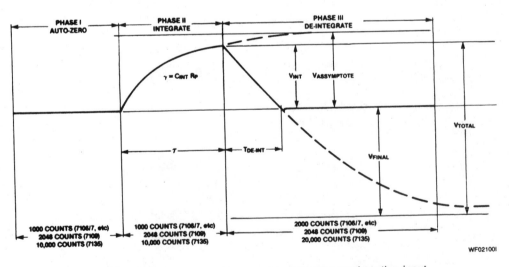

Integrator output waveforms with respect to integrator non-inverting input

Fig. 4-43B

(IN)

No-Ref-Cap Circuits

In many cases, the polarity of the input voltage is always known, or a different reference is required for different polarities. In other applications, it may be desirable for the reference voltage to vary in some manner during the conversion, for instance to achieve some nonlinear conversion function. In all these cases, the reference capacitor is undesirable or un-needed. Figure 4-44 shows the way in which it can be removed, and the desired reference voltage fed directly into the C_{REF} pin(s). Note that the reference source may be shorted out in some phases of the conversion, so a current-limiting impedance must be provided.

Low Noise Preamp Circuit

The noise performance of the ICL7106/7/9 family is controlled by the noise trapped on the auto-zero capacitor at the beginning of the integrate phase. This noise depends (in a complicated way) on the input noise of the buffer amplifier.

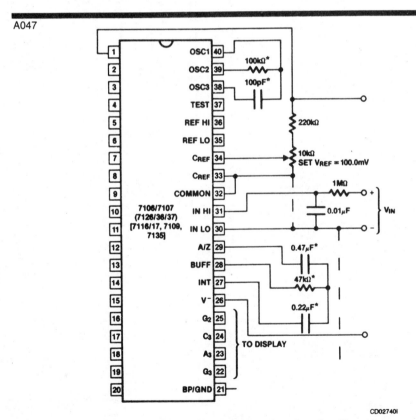

CD027401

*ICL7106/7 only. See data sheet for values for other parts.

A "No-ref-cap" circuit. This circuit gives correct
negative readings but very high (or O/R) readings for positive inputs

Fig. 4-44

If the built-in buffer is replaced by a low noise op amp, the noise performance could be improved, especially if gain can be introduced into this buffer, as can be done with the 2-chip devices. Figure 4-45 shows a way of doing this, the main losses being in higher input current and the lack of a true differential input. The switch network of the original is replaced by an ICL8053 driven in synchronism with the internal counter by using the original switch network-buffer combination, fed by resistive dividers, and a quad comparator to detect the various phases as shown.

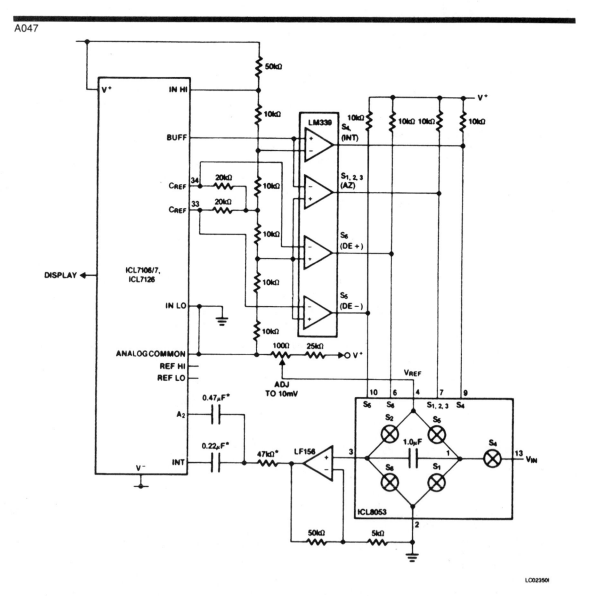

Fig. 4-45   Using an external low-noise preamp with the ICL7106/7/26   (IN)

Current I/P Converter Circuit

The normal voltage conversion circuits convert the differential input and reference voltages into corresponding currents in the integrating capacitor. In cases where the input signal is fundamentally a current in the appropriate range, it may make more sense to integrate it directly, rather than covert it to a voltage first, especially if the reference value is a current also.

The circuit of Fig. 4-46 injects the currents directly into the integrator input. The sources are switched in synchronism with the internal conversion phase by using the buffer amplifier and the voltage switching network in combination with a quad comparator, in the same manner as in Fig. 4-45. The normal auto-zero operation still occurs, provided the input is grounded during the appropriate phase.

A047

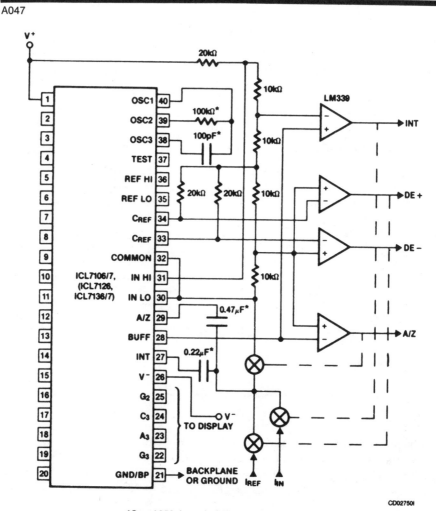

*See A052 for calculation on these values.

CD027501

Fig. 4-46   Current I/P converter circuit   (IN)

A047

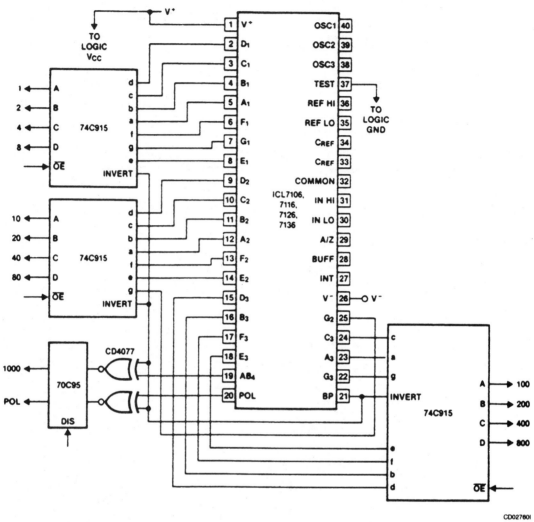

*For ICL7107/7117, tie "INVERT" high, use 0V supply for logic GND, and omit EX-NOR gates.

Fig. 4-47   BCD output from 7-segment O/P drive   (IN)

BCD Output from 7-Segment O/P Drive

Frequently it is necessary to provide a BCD output in addition to display driving, for peripheral output on panel meters, or for special decoding of upper/lower limit values, etc. The circuits shown here use a standard CMOS gate circuit to convert the 7-segment output of either LCD or LED drivers to BCD.

5
Digital and Microprocessor- Based Circuits

ICM7231 - ICM7234

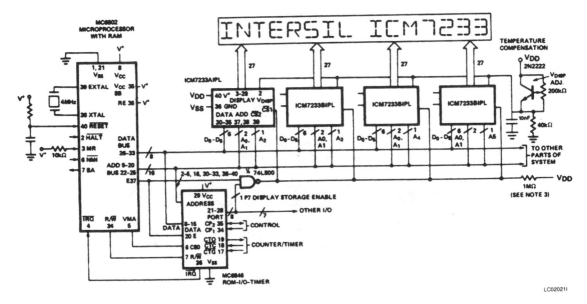

MC6802 microprocessor with 16 character 16 segment ASCII liquid crystal display.

Fig. 5-1

(IN)

ICM7231 - ICM7234

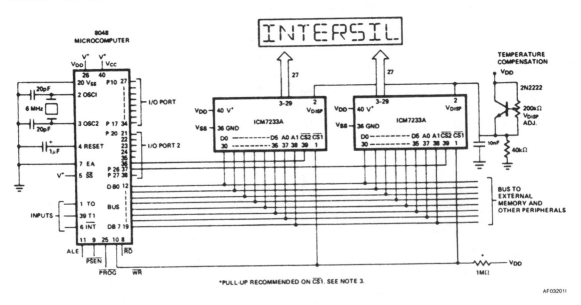

8048/IM80C48 microcomputer with 8 character 16 segment ASCII triplex liquid crystal display.

Fig. 5-2

(IN)

ICM7231 - ICM7234

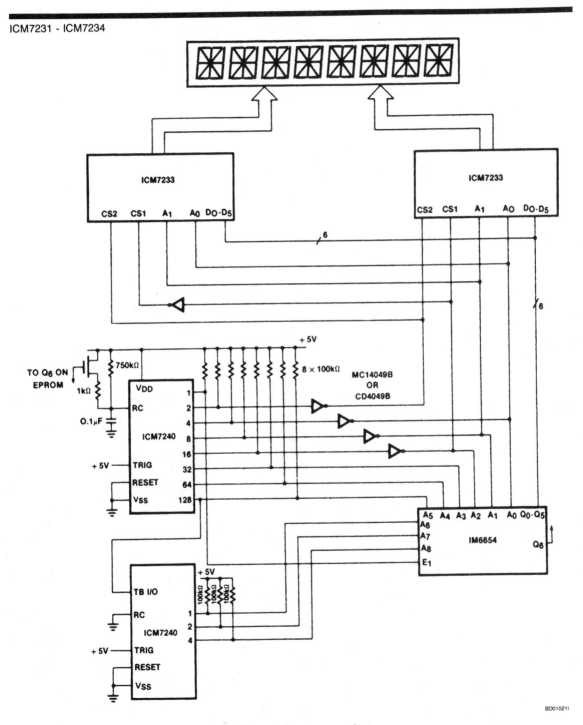

EPROM-coded message system.

Fig. 5-3

(IN)

ICM7231 - ICM7234

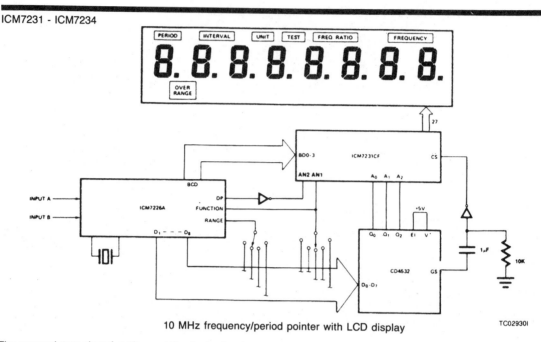

10 MHz frequency/period pointer with LCD display

TC02930I

The annunciators show function and the decimal points indicate the range of the current operation. The system can be efficiently battery operated.

Fig. 5-4

(IN)

ICM7231 - ICM7234

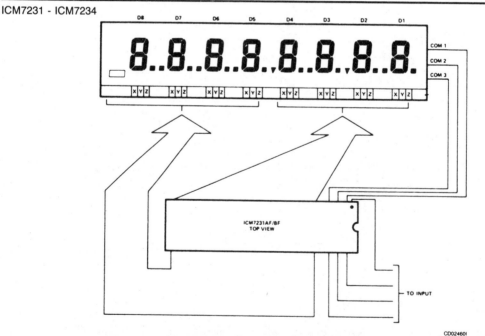

Fig. 5-5   "Forward" pin orientation and display connections

CD02460I

(IN)

ANALOG INPUT MICROPERIPHERAL

The following diagrams show interconnections of the MP20 (described in this data sheet) and also of Burr-Brown's MP10 analog output microperipheral (PDS-363) with Intel's 8080, National's SC/MP and Zilog's Z-80.

MP20

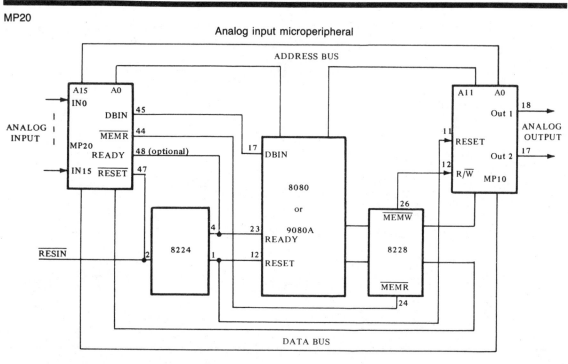

Fig. 5-6

MP10 and MP20 used with 8080

(BB)

MP20

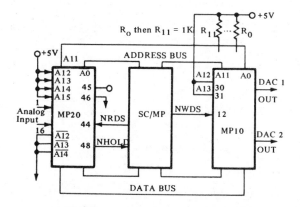

MP10 and MP20 used with the SC/MP

Fig. 5-7

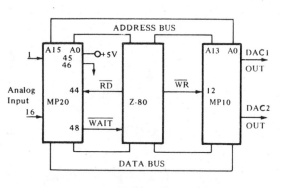

MP10 and MP20 used with the Z-80

(BB)

MP20

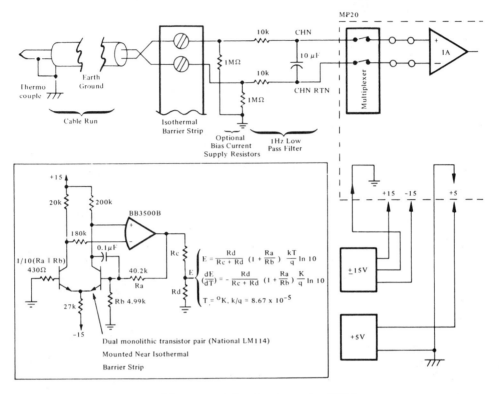

Thermocouple input system using MP20

Fig. 5-8   (BB)

MP20

	JUMPER		JUMPER
Single-ended Multiplexer	4 to 2; 4 to 77; 79 to 64; 15 to 14; 13 open	Address Bus (A0-A15)	Connect to 8080's address bus A0-A15
Differential Multiplexer	4 to 2; 77 to 79; 13 to 14; 15 open	Address Select ($\overline{A4}$-$\overline{A14}$)	Connect to +5V* or Ground
Amplifier	1 and 3 open for G = 2; R_{ext} between 1 and 3 for G≠2.	Control Bus	44 to 8228's $\overline{MEMR}$ output (pin 24) 45 to 8080's DBIN output (pin 17) 46 to ground 47 to 8224's $\overline{RESIN}$ input (pin 23) for normal operation 48 open for operation without halting CPU.
Input Range ±5V ±2.5V ±1.25V 0 - 5V 0 - 2.5V	65 to 63; 66 open; 67 to 68 65 to 63; 66 to 68; 67 open 65 to 63; 66 to 68; 63 to 67 65 to 64; 66 to 68; 67 open 65 to 64; 66 to 68; 63 to 67		
Output Coding	52 to 51 for binary; 52 to 50* for two's complement.	Data Bus (D0 - D7)	Connect to 8080's data bus.
* Through a 1kΩ resistor			

Fig. 5-9   (BB)

IM80C48, ICM7211M

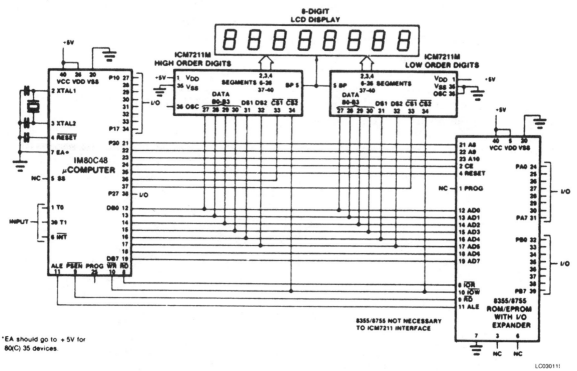

IM80C48 microprocessor interface

Fig. 5-10

(IN)

*EA should go to +5V for
80(C) 35 devices.

ICM7218

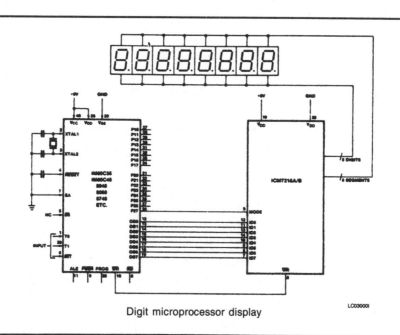

Digit microprocessor display

Fig. 5-11

(IN)

INTERRUPT POWERED MICROPROCESSOR

Power-up controlled by logic level functions is a popular digital hardware technique. The benefits of turning off logic functions may at first seem as a power conservation measure. Further consideration shows powered-down circuitry no longer responds to input signals, and will not produce false or spurious outputs to interfere with critical processing, resulting in lower error incidence.

The circuit illustrates the use of an interrupt signal to turn on a processor along with its support circuits. The TP0102N3 with VGS(th)2.4 Volt Max., turns on adequately with the drive available from 4001.

TP0102N3

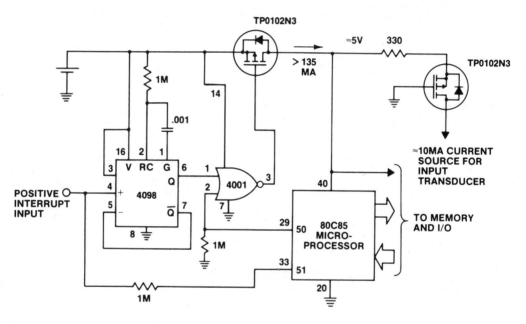

Fig. 5-12

(SU)

MICROSTEPPING MOTOR DRIVER

This microprocessor controlled stepper motor driver employs a high speed and efficient permanent-magnet motor with a normal 1.8 degree step to microstep at .028 degrees. Sine and cosine data from ROM-based table are alternately loaded into a register (sine section shown) and then coupled to the DAC-08 producing the reference voltage for the motor current control with the switching regulator. This regulator consists of a 3526 IC and a VN1210 DMOS device. Strobe pulse-width and V supply determine the motor speed. Sign-bit 7 determines the winding energized by Q2 and Q3. The cosine section is an exact mirror-image that powers motor winding 2 and 4.

DMOS advantages include excellent utilization of the 3526 output drive and simplified direct winding control from TTL circuits.

VN1210N5

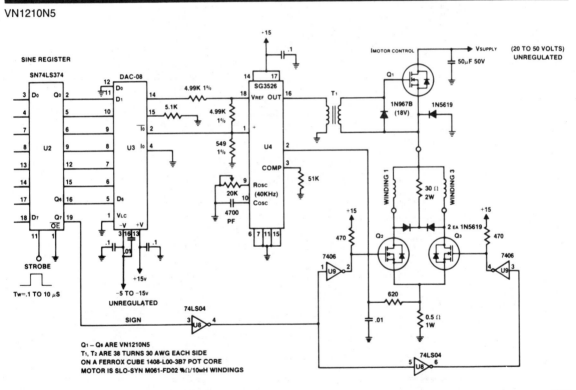

Fig. 5-13

(SU)

8-BIT MICROPROCESSOR INTERFACED ANALOG OUTPUT SYSTEM

MP10 ANALOG INPUT/OUTPUT

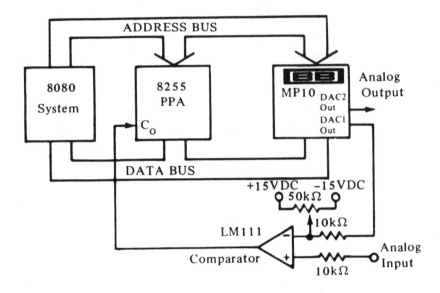

MP11 ANALOG INPUT/OUTPUT

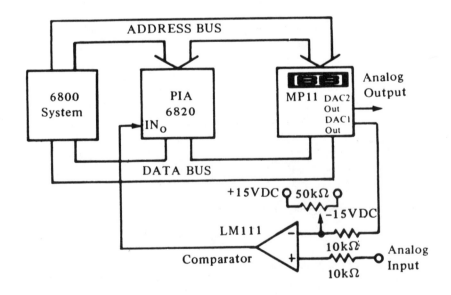

Fig. 5-14

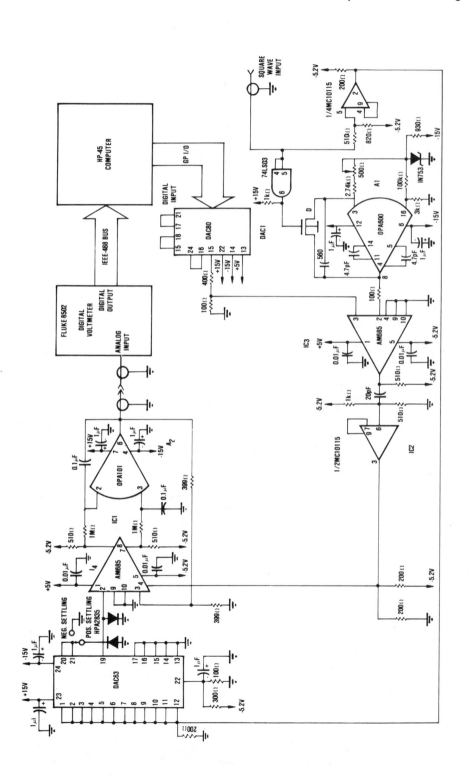

OPA101, OPA600

Computer controlled digitizer

Fig. 5-15

(BB)

ED-11, DC-7

ENCODER/DECODERS

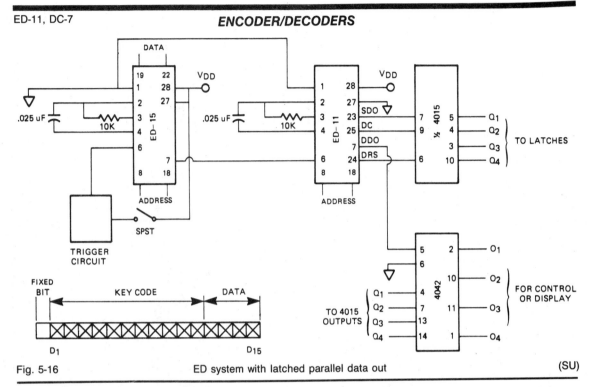

Fig. 5-16   ED system with latched parallel data out   (SU)

ED-11, DC-7

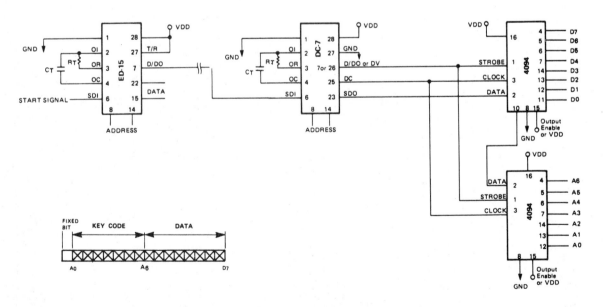

Fig. 5-17   DC-7 system with latched parallel data out   (SU)

ED-11, DC-7

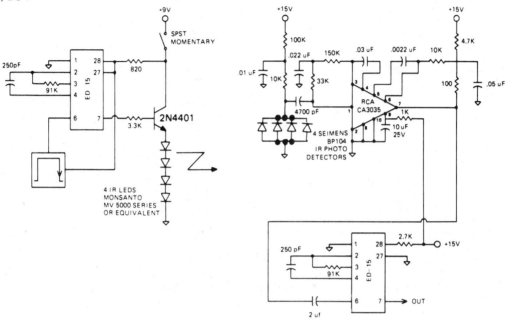

IR remote control transmitter/receiver

Fig. 5-18 (SU)

ED11, DC-7

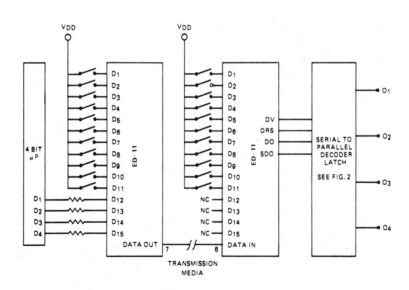

Block diagram showing basic configuration for transmitting
microprocessor data over remote control system using ED-11's as encoder/decoder.

Fig. 5-19 (SU)

ED-11, DC-7

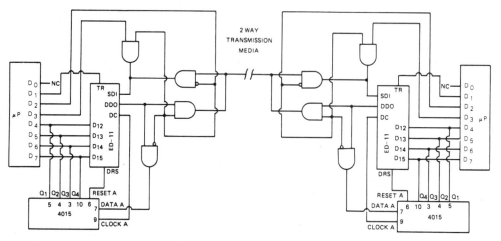

ED system illustrating "handshaking" capabilities of Supertex ED-11's

Fig. 5-20   (SU)

ED-11, DC-7

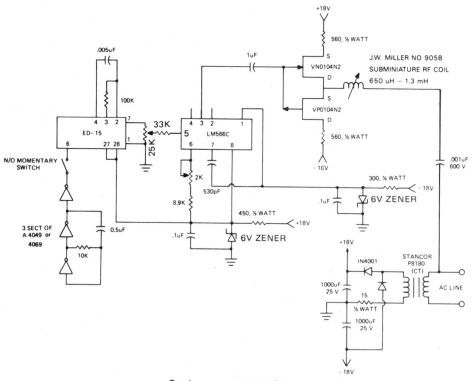

Carrier current transmitter

Fig. 5-21   (SU)

ED-11, DC-7

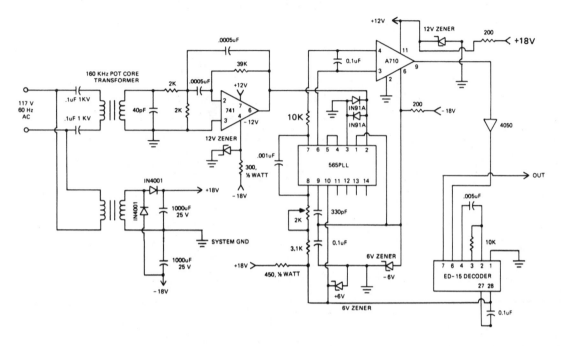

Carrier current receiver. 160 kHz transformer consists of an 18 × 11 mm ungapped pot core (Siemens, ferrocube, etc.) utilizing magnetics incorporated type "F" material wound with 80-½ turns of No. 35 wire for the secondary and 5-½ turns for the primary. This gives a turns ratio of approximately 15 to 1.

Fig. 5-22   (SU)

Cascading Encoder/Decorder Devices

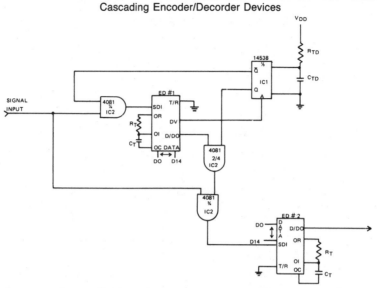

A simple means of cascading two ED devices to allow more than 1.07×10^9 addresses.

Fig. 5-23 (SU)

Cascading Encoder/Decorder Devices

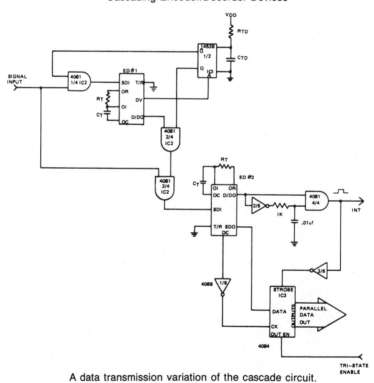

A data transmission variation of the cascade circuit.

Fig. 5-24 (SU)

PROGRAMMABLE DATA CODER

The DC-7A is a single monolithic chip using metal gate CMOS technology for low cost, low power, high yield and high reliability. This dual purpose circuit is capable of working either as an encoder or decoder on its own transmission in applications where exclusive recognition of address codes are required in addition to transmission or reception of 8 Data Bits. It will decode 1 for 128 address codes. In the transmit mode this circuit is capable of generating the possible codes by connecting the Address and Data Inputs to VDD or GND for a "1" or a "0." In the receive mode this circuit is capable of decoding the transmitted signals and simultaneously making comparisons to the local address codes for identification.

DC-7A

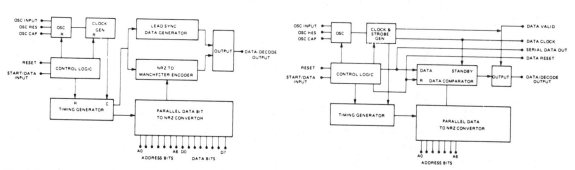

Fig. 5-25

(SU)

DC-7A

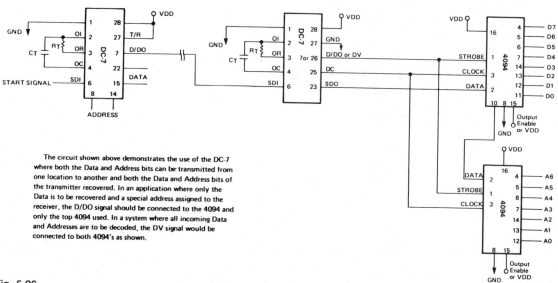

The circuit shown above demonstrates the use of the DC-7 where both the Data and Address bits can be transmitted from one location to another and both the Data and Address bits of the transmitter recovered. In an application where only the Data is to be recovered and a special address assigned to the receiver, the D/DO signal should be connected to the 4094 and only the top 4094 used. In a system where all incoming Data and Addresses are to be decoded, the DV signal would be connected to both 4094's as shown.

Fig. 5-26

(SU)

DIGITAL SIGNAL PROCESSOR

The MS2014 is a real time general purpose digital signal processor (DSP) which is easily programmed to perform digital filtering and level detection. The architecture of the FAD comprises a cascadable second order recursive filter and level detector using dedicated multipliers, adders and delay elements.

The data controlling the response of the MS2014 is stored in an external PROM or RAM and consists of a list of filter coefficients and comparison levels. This simple data format means that the user does not need an expensive development system at the design stage (in contrast to other DSP devices, which use microprocessor-based structures and require considerable software development effort to realize their function). The off-chip data memory allows for easy adaptive control, even when complicated algorithms are to be implemented.

The filter and detector have been designed to give maximum flexibility in use and can easily generate most of the functions required in tone detector, spectral analysis, adaptive filter and speech synthesis systems.

Features

☐ Linear 16-Bit Data
☐ 13-Bit Coefficient
☐ 2 MHz Operating Clock Frequency
☐ Serial Operation
☐ 448 Bits of On-Chip Shift Register Data Storage for 8th Order Multiplex
☐ Nth Order Multiplexing (N ≤ 8)
☐ TTL Compatible
☐ Single +5 V Supply

Applications

☐ Low Cost Digital Filtering
☐ Level Detection
☐ Spectral Analysis
☐ Tone Detectors (Multi-Frequency Receivers)
☐ Speech Synthesis and Analysis
☐ Data Modems
☐ Group Delay Equalisers (All-Pass Networks)

MS2014

Pin connections - top view

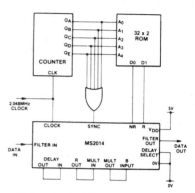

2nd order 32 kHz bandwidth filter

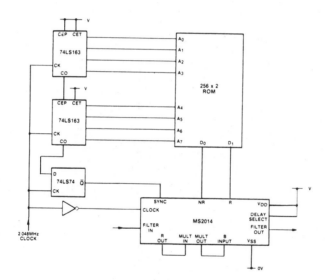

16th order 4 kHz bandwidth filter

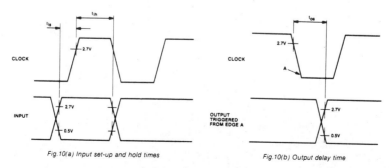

Fig.10(a) Input set-up and hold times

Fig.10(b) Output delay time

Input and output timing

Fig. 5-27

(PL)

SP1450, SP1455

PCM SIGNAL MONITOR CIRCUIT

Characteristic	Pin	Value			Units	Conditions
		Min.	Typ.	Max.		
Max. Input Frequency						
SP1450	13		—	25.5	M band/s	See note 1 below
SP1455	13		—	105	M band/s	
Stretched output pulse width	15	0.5	0.7	2	μS	c_1 390 pF $R_1 = 27$ kΩ using circuit of Fig. 7 (see note 2 below)
Error pulse width SP1455	13	4.25	—	5.25	nS	Input freq. 105 M band/s
Error pulse amplitude	13	300	—	—	mV	At max input frequency
Spurious pulse amplitude	13	—	—	50	mV	At max. input frequency

NOTE 1: These figures are the max.input symbol rates. For 4B3T codes, the effective bit rate is 4/3 x (input frequency).
NOTE 2: Resistor and capacitor values quoted are absolute values; temperature coefficients and tolerances have not been taken into account.

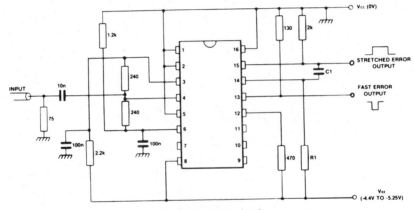

Functional test circuit

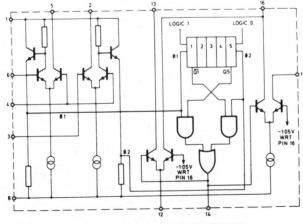

Fig. 5-28   Circuit diagram of SP1450/SP1455   (PL)

6
Optoelectronic Circuits

OPTICAL COUPLING (GENERAL CONFIGURATIONS)

Compensated optical couplers use negative feedback to cancel the errors in linearity and gain of two matched couplers. The compensated circuits that are shown are a current-to-current configuration and a voltage-to-voltage configuration.

Differential optical coupling provides superior performance and makes a major advance in linearity and stability.

A very simple implementation of differential optical coupling using one LED, two photodiodes, and two op amps for both unidirectional (a) and bidirectional (b) operation is shown. Unidirectional operation limits amplification to positive signals only, whereas bidirectional handles both positive and negative signals by providing a bias current source for the photodiodes.

Optical Coupling

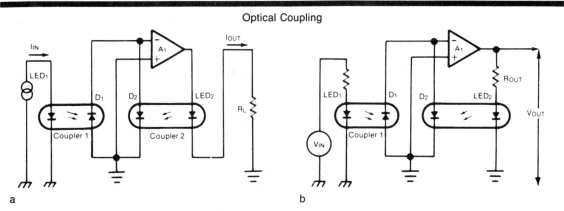

Compensated circuits. Negative feedback cancels errors in linearity and gain of the two matched couplers. The circuit shown in (a) operates in a current mode, but adding two resistors (b) allows it to function in a voltage-to-voltage mode.

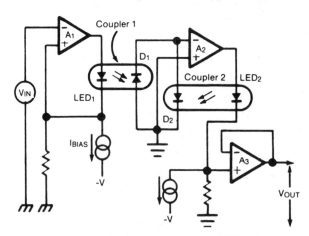

Improved linearity. Nonlinearity of optical couplers is lowest when the LED is driven from a current source. In this circuit both LEDs are biased from separate current sources, and an operational amplifier serves as a voltage-to-current converter.

Fig. 6-1

(BB)

Optical coupling (general configurations)

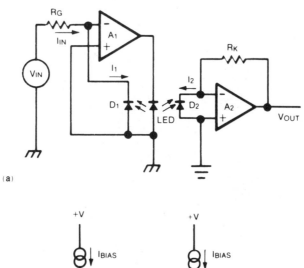

(a)

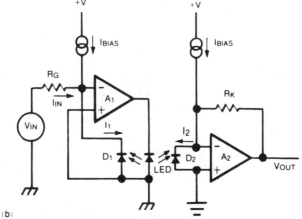

(b)

Better performance. Additional photodiode matched to the primary light source's sensor diode can virtually eliminate nonlinearity and temperature drift errors of an optical coupler. The circuit of (a) is for unidirectional operation; the one in (b) is bidirectional.

Fig. 6-2   (BB)

OPA2111

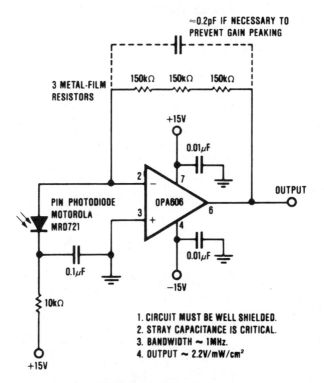

High-speed photodetector.

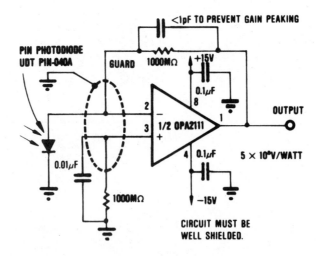

Sensitive photodiode amplifier.

Fig. 6-3   (BB)

ICM7217/ICM7227

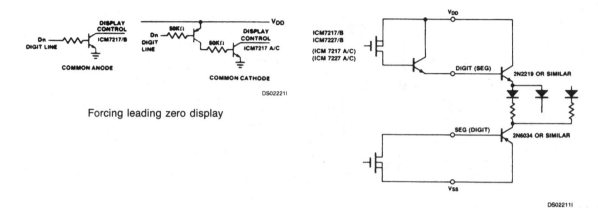

Forcing leading zero display

Driving high current displays

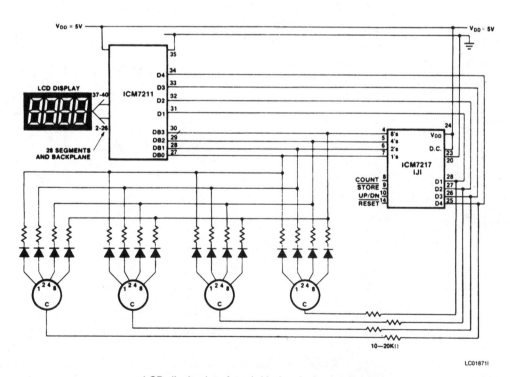

LCD display interface (with thumbwheel switches)

Fig. 6-4

ICM7217/ICM7227

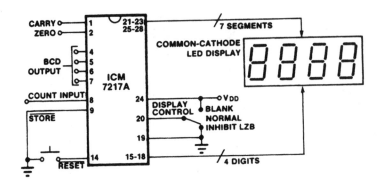

Unit counter

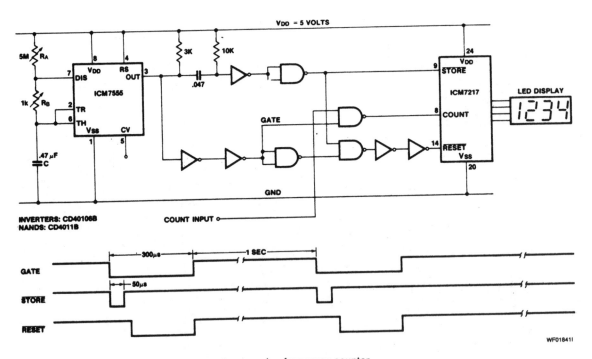

Inexpensive frequency counter

WF01841I

Fig. 6-5

(IN)

ICM7217/ICM7227

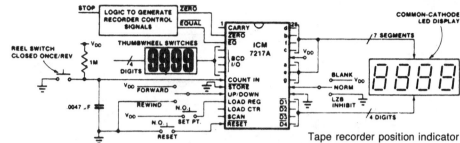

Fig. 6-6   Tape recorder position indicator

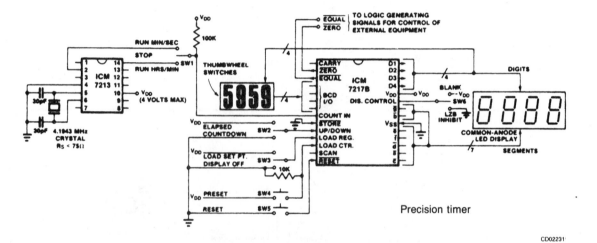

Precision timer

CD02231

ICM7217/ICM7227

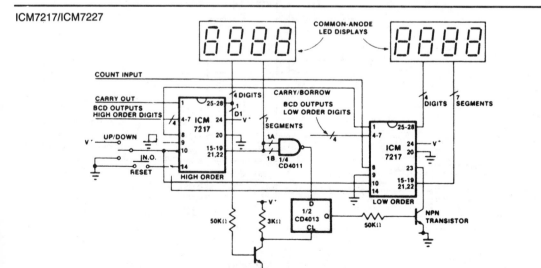

Fig. 6-7   CD02240I

ICM7217/ICM7227

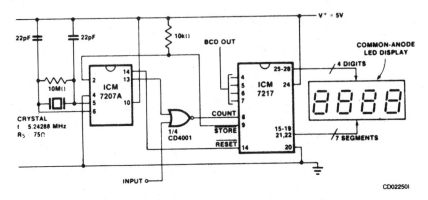

Precision frequency counter (MHz maximum)

Fig. 6-8

ICM7217/ICM7227

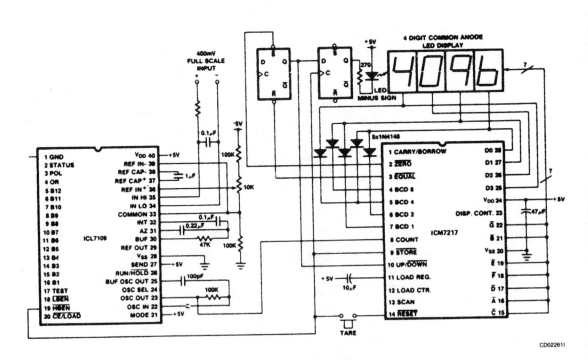

Auto-tare system for A/D converter

Fig. 6-9

ICM7243

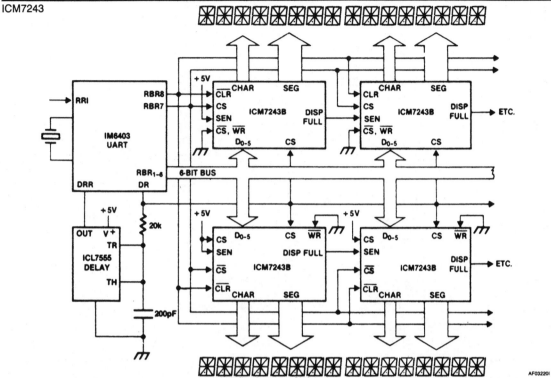

UART converts data stream to parallel bytes. Bit 7 of each word sets which row data will be entered into. Bit 8 will blank and reset whole display if low. Each MODE pin should be tied high. ICM7243A can also be used, with inverter on RBR7 for one row.

Fig. 6-10   Driving two rows of characters from a serial input.   (IN)

ICM7243

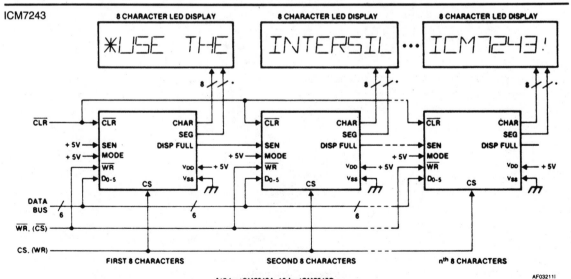

*17 for ICM7243A, 15 for ICM7243B

Fig. 6-11   Multicharacter display using serial access mode   (IN)

ICM7236

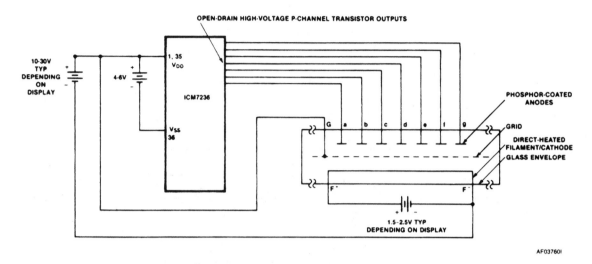

Typical dc vacuum fluorescent display connection

Fig. 6-12   (IN)

ICM7211

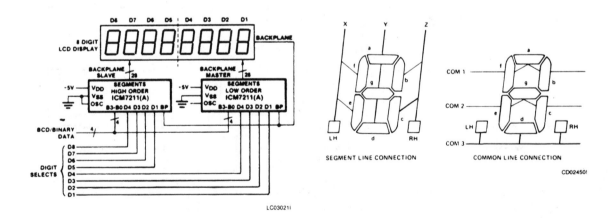

Ganged ICM7211's driving 8-digit LCD display   Connection diagrams for typical 7-segment displays

For other ICM7211 circuits see Chapter 3.

Fig. 6-13   (IN)

ICM7243

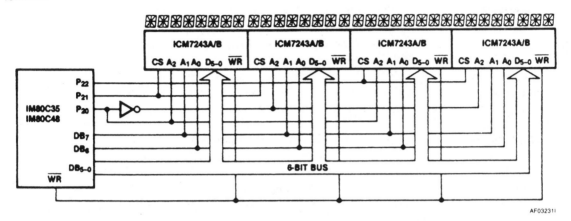

AF03231I

Random Access 32-Character Display in IM80C48 system.

One port line controls A$_2$, other two are CS lines. 8-bit data bus drives 6 data and 2 address lines. MODE should be GrouNDed on each part.

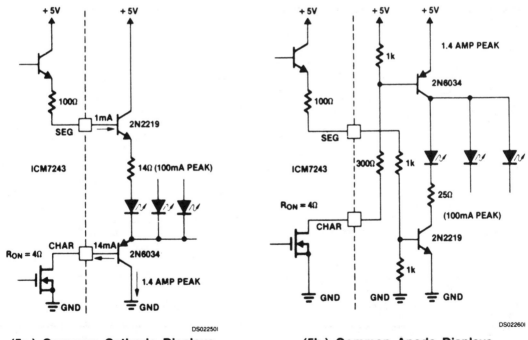

DS02250I

(5a.) Common Cathode Displays

DS02260I

(5b.) Common Anode Displays

Fig. 6-14   Driving large displays   (IN)

DC VACUUM FLUORESCENT DISPLAY

Each device in the ICM7235 family provides signals for directly driving the anode terminals of a four-digit, 7-segment non-multiplexed vacuum fluorescent display. The outputs are taken from the drains of high-voltage, low-leakage p-channel FETs. Each is capable of withstanding > -35 V with respect to VDD. In addition, the inclusion of an $\overline{ON}$/OFF input allows the user to disable all segments by connecting pin 5 to VDD; this same input may also be used as a brightness control by applying a signal swinging between VDD and VSS and varying its duty cycle.

The ICM7235 may also be used to drive nonmultiplexed common cathode LED displays by connecting each segment output to its corresponding display input, and tying the common cathode to VSS. Using a power supply of 5 V and an LED with a forward drop of 1.7 V results in an "ON" segment current of about 3 mA, enough to provide sufficient brightness for displays of up to 0.3" character height.

Note that these devices have two VDD terminals, and each should be connected to the positive supply voltage. This double connection is necessary to minimize the effects of bond wire resistance, which could be a problem due to the high display currents.

ICM7235

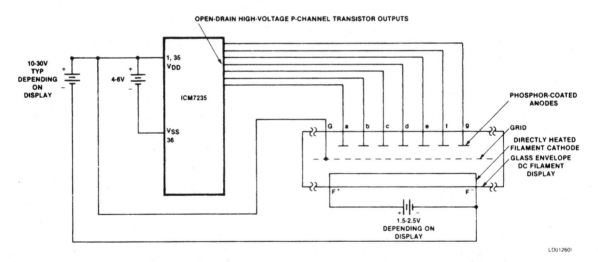

Fig. 6-15   ICM7235 Typical dc vacuum fluorescent display connection   (IN)

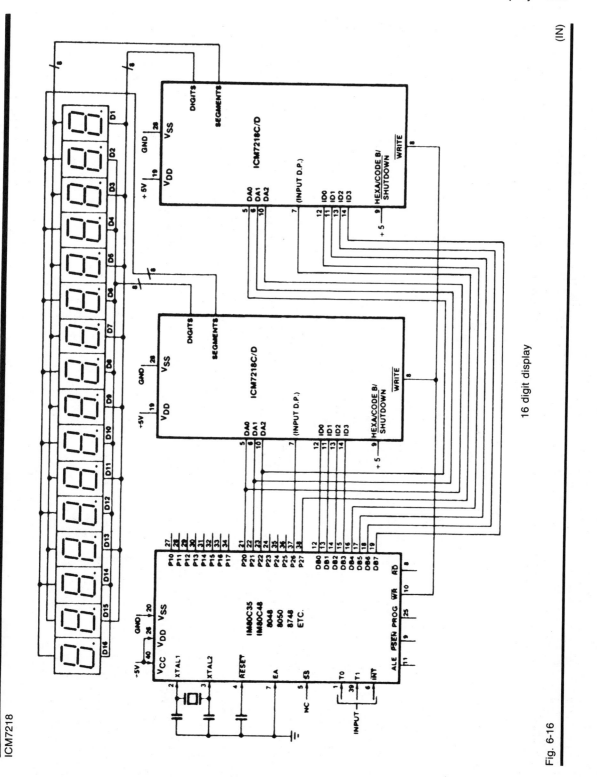

16 digit display

ICM7218

Fig. 6-16

ISOLATED 0-300 V SYSTEM SUPPLY

VN03's low power gate drive permits the voltage generating optocoupler shown in this circuit to provide gate control, H.V. isolation, and serve in the current limit scheme with the 33 ohms sense resistor. This type of isolator should find additional service wherever isolation and MOS transistors are used to handle dc, and lower frequency ac signals.

VN0340N1, VN0345N1

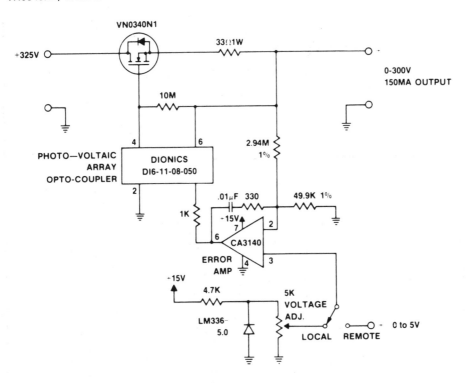

Fig. 6-17

GATED PRECISION OPTICAL SOURCE

This circuit rejects variations in supply voltage and LED conduction to provide constant optical output. Two VN02 transistors serve as modulator and regulator for this precision optical source. Q4 senses Q3's optical output to produce a voltage error signal for current source control. (The photo-transistor emitter, when connected to a signal detection amplifier, produces a full duplex system capable of fiber-optic coupling.) Laser-diode printing heads, where a matrice of LEDs require optical regulation, could gain from the compact Quad VN02 version.

(SU) VN0206N3, VN0206N5

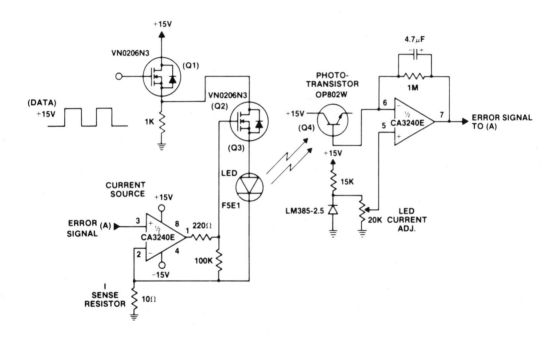

Fig. 6-18

(SU)

7
Audio and
Radio Circuits

I-F AMPLIFIER AND DETECTOR

The SL6700C is a single or double conversion i-f amplifier and detector for AM radio applications. Its low power consumption makes it ideal for hand held applications. Normally the SL6700C will be fed with a first i-f signal of 10.7 MHz or 21.4 MHz; there is a mixer for conversion to the first or second i-f, a detector, an AGC generator with optional delayed output and a noise blanker monostable.

Features

☐ High Sensitivity: 10 μV minimum
☐ Low Power: 8 mA Typical at 6 V
☐ Linear Detector

Applications

☐ Low Power AM/SSB Receivers

PCM53

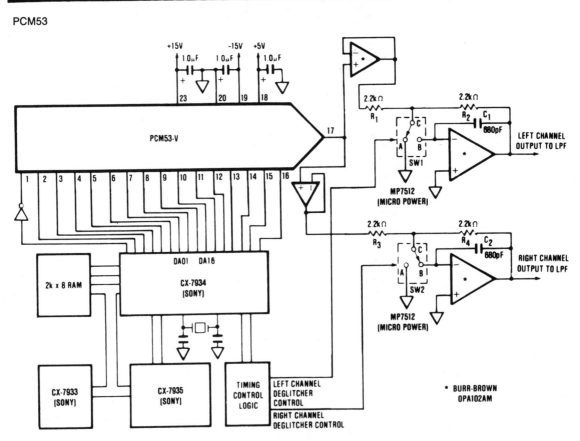

A single PCM52/53 used to obtain both left and right channel output in a typical digital audio system.

Fig. 7-1

(BB)

SL6700C

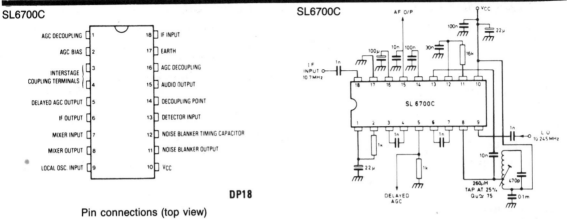

Pin connections (top view)

DP18

Fig. 7-2

AM double conversion receiver with noise blanker

(PL)

SL6700C

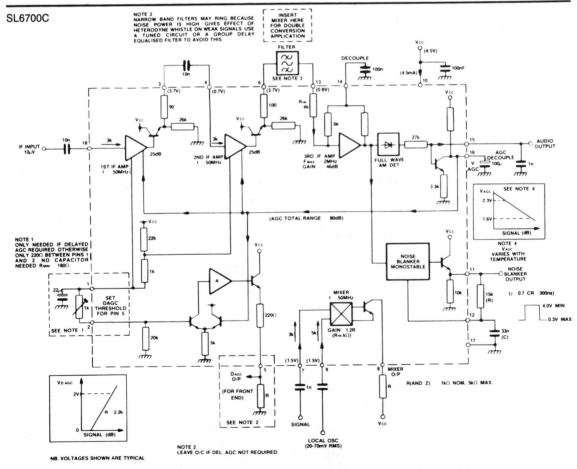

Fig. 7-3

SL6700C typical application circuit showing interfacing

(PL)

SINGLE CHIP MIXER/OSCILLATOR, I-F AMPLIFIER/DETECTOR

The SL6653 is a complete single chip mixer/oscillator, i-f amplifier and detector for FM cellular radio, cordless telephones and low power radio applications. Supply current is less than 2 mA from a supply voltage in the range 2.5 V to 7.5 V.

The SL6653 affords maximum flexibility in design and use. It is supplied in a dual-in-line hermetic package.

Features

- ☐ Low Power Consumption (1.5 mA)
- ☐ Single Chip Solution
- ☐ Guaranteed 100 MHz Operation

Quick Reference Data

- ☐ Supply voltage 2.5 V to 7.5 V
- ☐ Sensitivity 3 μV

Applications

- ☐ Mobile Radio Telephones
- ☐ Cordless Telephones

Absolute Maximum Ratings

Supply voltage	10 V
Storage temperature	$-55°$ C to $+150°$ C
Operating temperature	$-55°$ C to $+125°$ C
Mixer input	1 V rms

SL6653

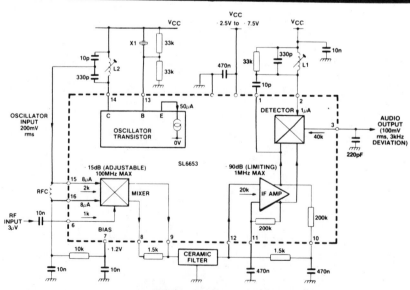

Fig. 7-4   Functional diagram   (PL)

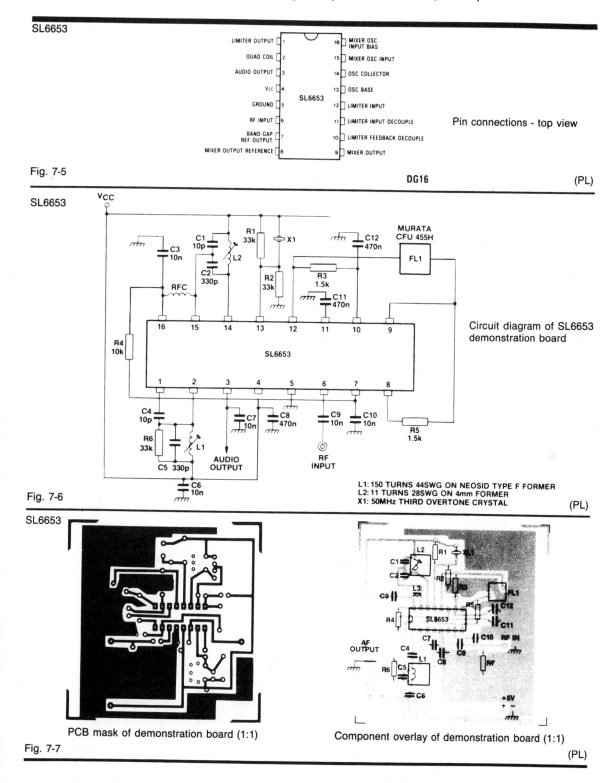

SL6653

Pin connections - top view

LIMITER OUTPUT — 1	16 — MIXER OSC INPUT BIAS
QUAD COIL — 2	15 — MIXER OSC INPUT
AUDIO OUTPUT — 3	14 — OSC COLLECTOR
Vcc — 4	13 — OSC BASE
GROUND — 5	12 — LIMITER INPUT
RF INPUT — 6	11 — LIMITER INPUT DECOUPLE
BAND-GAP REF OUTPUT — 7	10 — LIMITER FEEDBACK DECOUPLE
MIXER OUTPUT REFERENCE — 8	9 — MIXER OUTPUT

Fig. 7-5

DG16 (PL)

SL6653

Circuit diagram of SL6653
demonstration board

L1: 150 TURNS 44SWG ON NEOSID TYPE F FORMER
L2: 11 TURNS 28SWG ON 4mm FORMER
X1: 50MHz THIRD OVERTONE CRYSTAL

Fig. 7-6 (PL)

SL6653

PCB mask of demonstration board (1:1)

Component overlay of demonstration board (1:1)

Fig. 7-7 (PL)

LOW POWER IF/AF CIRCUIT FOR FM CELLULAR RADIO

The SL6652 is a complete single chip mixer/oscillator, i-f amplifier and detector for FM cellular radio, cordless telephones and low power radio applications. It features an exceptionally stable RSSI (Received Signal Strength Indicator) output using a unique system of detection. Supply current is less than 2 mA from a supply voltage in the range 2.5 V to 7.5 V.

Features

□ Low Power Consumption (1.5 mA)
□ Single Chip Solution
□ Guaranteed 100 MHz Operation
□ Exceptionally Stable RSSI

Applications

□ Cellular Radio Telephones
□ Cordless Telephones

Quick Reference Data

□ Supply Voltage 2.5 V to 7.5 V
□ Sensitivity 3 μV
□ Co-Channel Rejection 7 dB

The SL6652 is a very low power, high performance integrated circuit intended for i-f amplification and demodulation in FM radio receivers. It comprises:

□ A mixer stage for use up to 100 MHz
□ An uncommitted transistor for use as an oscillator
□ A current sink for biasing this transistor
□ A limiting amplifier operating up to 1.5 MHz
□ A quadrature detector with differential AF output
□ An RSSI (Received Signal Strength Indicator) output

Mixer. The mixer is single balanced with an active load. Gain is set externally by the load resistor although the value is normally determined by that required for matching into the ceramic filter. It is possible to use a tuned circuit but an increase in mixer gain will result in a corresponding reduction of the mixer input intercept point.

The RF input is a diode-biased transistor with a bias current of typically 300 μA. The oscillator input is differential but would normally be driven single-ended. Special care should be taken to avoid accidental overload of the oscillator input.

Oscillator. The oscillator consists of an uncommitted transistor and a separate current sink. The user should ensure that the design of oscillator is suitable for the type of crystal and frequency required; it may not always be adequate to duplicate the design shown in this data sheet.

I-f amplifier. The limiting amplifier is capable of operation to at least 1 MHz and the input impedance is set by an external resistor to match the ceramic filter. Because of the high gain, pins 12 and 13 must be adequately bypassed.

Detector. A conventional quadrature detector is fed internally from the i-f amplifier; the quadrature input is fed externally using an appropriate capacitor and phase shift network. A differential output is provided to feed a comparator for digital use, although it can also be used to provide AFC.

RSSI output. The RSSI output is a current source with value proportional to the logarithm of the i-f input signal amplitude. There is a small residual current due to noise within the amplifier (and mixer) but beyond this point there is a measured and guaranteed 70 dB dynamic range. The typical range extends to 92 dB, independent of frequency, and with exceptionally good temperature and supply voltage stability.

Supply voltage. The SL6652 will operate reliably from 2.5 V to 7.5 V. The supply line must be decoupled with 470 nF using short leads.

Internal bias voltage. The internal band gap reference must be externally decoupled. It can be used as an external reference but must not be loaded heavily; the output impedance is typically 14 ohms.

SL6652

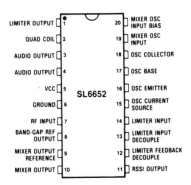

DG20

Fig. 7-8

(PL)

SL6652

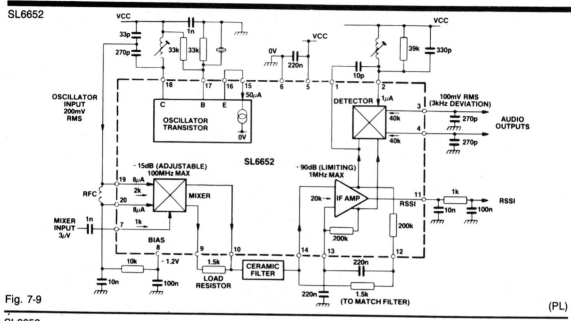

Fig. 7-9

(PL)

SL6652

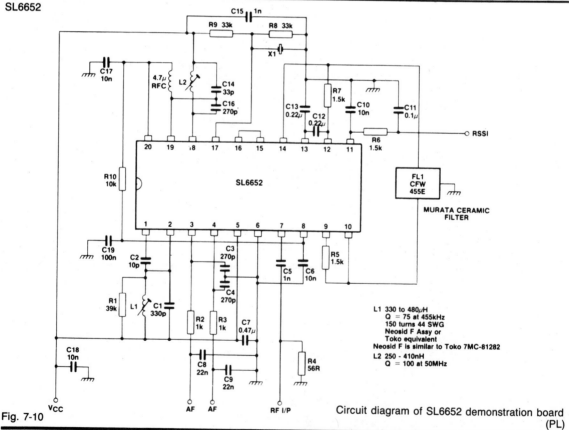

L1 330 to 480μH
Q = 75 at 455kHz
150 turns 44 SWG
Neosid F Assy or
Toko equivalent
Neosid F is similar to Toko 7MC-81282
L2 250 - 410nH
Q = 100 at 50MHz

Circuit diagram of SL6652 demonstration board
(PL)

Fig. 7-10

SL6652

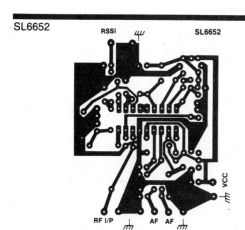

RSSI   SL6652

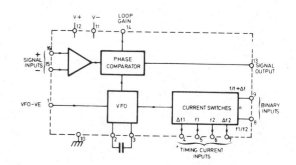

RSSI

PCB mask of demonstration board (1:1)

Component overlay of demonstration board (1:1)

Fig. 7-11

(PL)

SL652C

MODULATOR/VARIABLE FREQUENCY OSCILLATOR

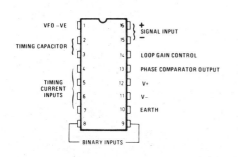

Pin connections (top view)

SL652C block diagram

Fig. 7-12

(PL)

SL652C   Modulator/phase-locked loop variable frequency oscillator

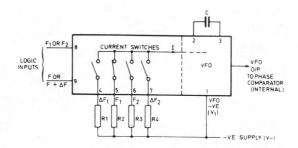

VFO and binary interface

Fig. 7-13

(PL)

SL652C

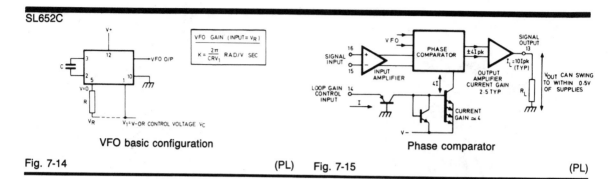

VFO basic configuration

Phase comparator

Fig. 7-14

(PL)   Fig. 7-15

(PL)

8
Alarms and Safety/ Security Circuits

CMOS PHOTO-ELECTRIC SMOKE DETECTOR INTEGRATED CIRCUIT

This low power CMOS circuit is intended for use in a pulsed LED/silicon cell smoke detector system. It is designed for use in low power, battery operated, consumer applications with a minimum of external components. This device meets UL217 requirements and is available in a 16-pin plastic DIP.

SD2

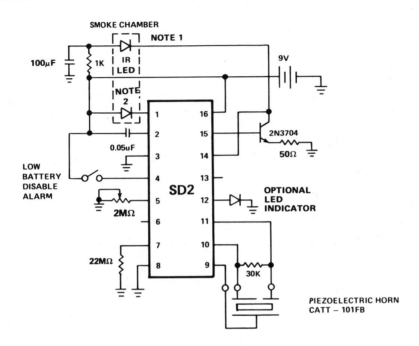

Notes: 1. IR Diode RCA Type SG 1010A or Spectronics Type SE 5455-4
Clairex Type CLED-1

2. IR Photo detectors Vactec VTS4085

Fig. 8-1   (SU)

IONIZATION CHAMBER TYPE SMOKE DETECTOR CIRCUIT

The SD3A is a CMOS integrated circuit designed for an ionization chamber type smoke detector that directly drives a piezo electric horn. It satisfies UL217 requirements and is available in a 14-lead plastic DIP.

Designed and built for an efficient, low component count, smoke detector system, the SD3A has numerous features that allow increased alarm effectiveness and reduced false triggering. With an improved offset voltage and built-in hysteresis this device requires less ion source and has increased sensitivity.

The horn output of this circuit can be a continuous or intermittent alarm. An optional LED indicator can be used to monitor the battery level. The SD3A operates on a single 9-volt alkaline or zinc carbon battery. It also may be used in multiple station connection applications.

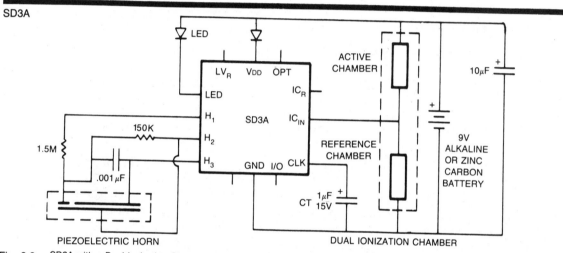

Fig. 8-2   SD3A with a Dual Ionization Chamber and Piezoelectric Horn together with an LED as battery connection indicator   (SU)

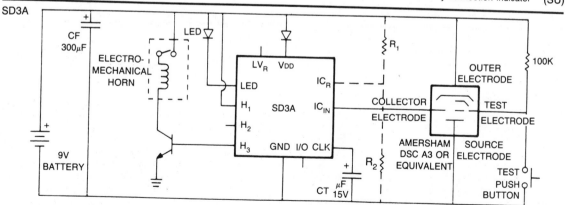

SD3A   SD3A with an Amersham DSC A3 concentric chamber and an electromechanical horn. Special features are optional R1/R2 resistor network for adjusting comparator trip voltage and built-in-test electrode for in-circuit alarm test.

Fig. 8-3

(SU)

SD3A

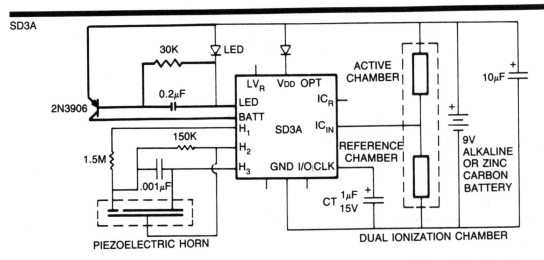

SD3A with dual ion chamber, piezoelectric horn, LED, battery impedance check, and intermittent horn.

Fig. 8-4   (SU)

SD3A

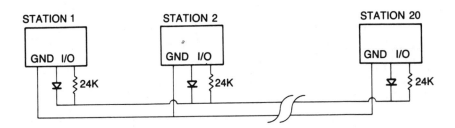

Fig. 8-5   (PU)

TDE1767, TDE1787

TYPICAL APPLICATIONS

OPEN LOAD DETECTION

$$\frac{V_{CC}\, R_1}{R_2} \geqslant 50\, mV$$

$$R_S\, I_L \geqslant 100\, mV$$

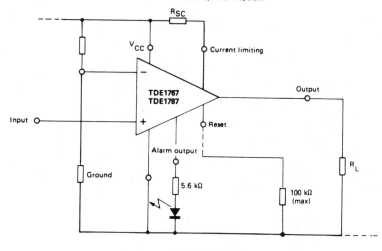

DRIVING LAMPS, RELAYS, etc...

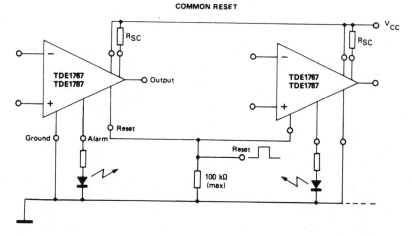

COMMON RESET

Fig. 8-6

(TH)

TDE1767, TDE1787

PARALLEL DRIVING OF LOADS UP TO 1 A

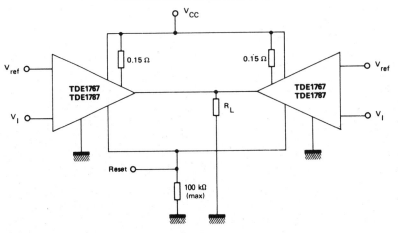

USING ALARM OUTPUT

PARALLEL ALARM OUTPUTS

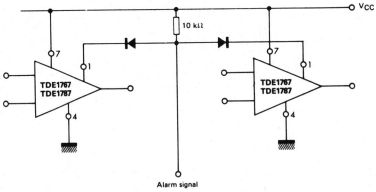

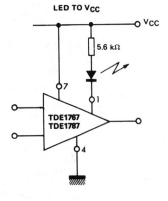

LED TO V_CC

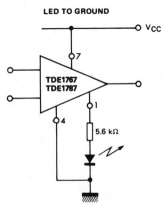

LED TO GROUND

Fig. 8-7

(TH)

INTERFACE BETWEEN HIGH VOLTAGE AND LOW VOLTAGE SYSTEMS

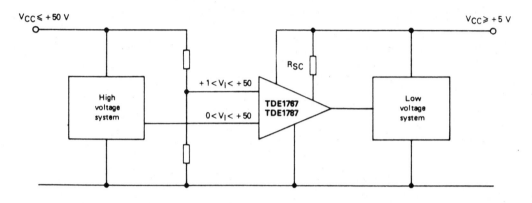

INCREASING OUTPUT CURRENT UP TO 10 A

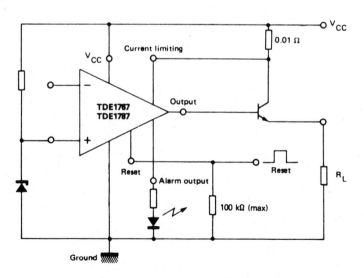

Fig. 8-8

9
Special-Purpose Circuits

MV5089   Dual-tone multifrequency generator

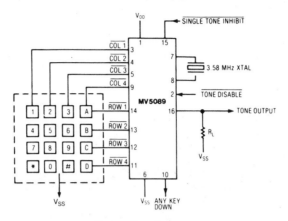

Fig. 9-1   Connection diagram   (PL)

MV8860

Dual-tone multifrequency decoder

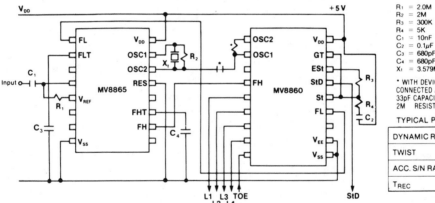

R_1 = 2.0M
R_2 = 2M
R_3 = 300K
R_4 = 5K
C_1 = 10nF
C_2 = 0.1μF
C_3 = 680pF
C_4 = 680pF
X_1 = 3.579MHz

* WITH DEVICE POWER SUPPLY CONNECTED AS FIG. 5B, 33pF CAPACITOR AND 2M RESISTOR REQUIRED

TYPICAL PERFORMANCE

DYNAMIC RANGE	30dB
TWIST	± 10dB
ACC. S/N RATIO	14dB
T_{REC}	30ms

Fig. 9-2   Single-ended input receiver using the MV8865 (5 V operation)   (PL)

MV8860

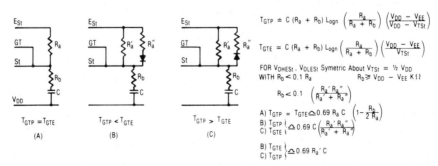

$$T_{GTP} = C (R_a + R_b) \, L_{ogn} \left(\frac{R_a}{R_a + R_b} \right) \left(\frac{V_{DD} - V_{EE}}{V_{DD} - V_{TSt}} \right)$$

$$T_{GTE} = C (R_a + R_b) \, L_{ogn} \left(\frac{R_a}{R_a + R_b} \right) \left(\frac{V_{DD} - V_{EE}}{V_{TSt}} \right)$$

FOR V_{OHESt}, V_{OLESt} Symetric About $V_{TSt} = \frac{1}{2} V_{DD}$
WITH $R_b < 0.1 \, R_a$ $R_b \gtrless V_{DD} - V_{EE} \, K\Omega$

$$R_b < 0.1 \left(R_a \frac{R_a' R_a''}{R_a' + R_a''} \right)$$

A) $T_{GTP} = T_{GTE} \triangleq 0.69 \, R_a \, C \left(1 - \frac{R_b}{2 \, R_a} \right)$
B) T_{GTP}
C) T_{GTE} } $\triangleq 0.69 \, C \left(\frac{R_a' R_a''}{R_a' + R_a''} \right)$

B) T_{GTE}
C) T_{GTP} } $\triangleq 0.69 \, R_a' \, C$

Fig. 9-3   Guard time adjustment   (PL)

MV8860

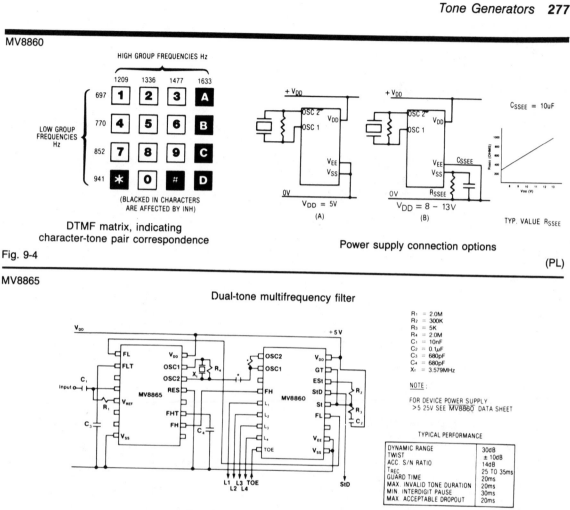

HIGH GROUP FREQUENCIES Hz

DTMF matrix, indicating character-tone pair correspondence

(BLACKED IN CHARACTERS ARE AFFECTED BY INH)

Power supply connection options

$C_{SSEE} = 10uF$

$V_{DD} = 5V$ (A)

$V_{DD} = 8 - 13V$ (B)

TYP. VALUE R_{SSEE}

Fig. 9-4   (PL)

MV8865

Dual-tone multifrequency filter

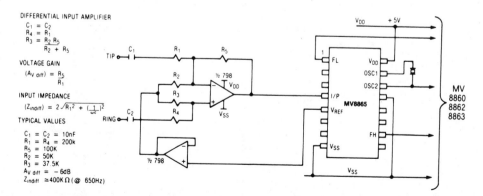

$R_1 = 2.0M$
$R_2 = 300K$
$R_3 = 5K$
$R_4 = 2.0M$
$C_1 = 10nF$
$C_2 = 0.1\mu F$
$C_3 = 680pF$
$C_4 = 680pF$
$X_1 = 3.579MHz$

NOTE:

FOR DEVICE POWER SUPPLY
>5 25V SEE MV8860 DATA SHEET

TYPICAL PERFORMANCE

DYNAMIC RANGE	30dB
TWIST	± 10dB
ACC. S/N RATIO	14dB
T_{REC}	25 TO 35ms
GUARD TIME	20ms
MAX. INVALID TONE DURATION	20ms
MIN. INTERDIGIT PAUSE	30ms
MAX. ACCEPTABLE DROPOUT	20ms

Fig. 9-5   Single-ended input receiver using the MV8860 (5 V operation)

(PL)

MV8865

DIFFERENTIAL INPUT AMPLIFIER

$C_1 = C_2$
$R_4 = R_1$
$R_3 = \dfrac{R_2 R_5}{R_2 + R_5}$

VOLTAGE GAIN

$(A_{V\,diff}) = \dfrac{R_5}{R_1}$

INPUT IMPEDANCE

$(Z_{indiff}) = 2\sqrt{R_1^2 + (\dfrac{1}{\omega c})^2}$

TYPICAL VALUES

$C_1 = C_2 = 10nF$
$R_1 = R_4 = 200k$
$R_5 = 100K$
$R_2 = 50K$
$R_3 = 37.5K$
$A_{V\,diff} = -6dB$
$Z_{indiff} \approx 400K\Omega\,(@\ 650Hz)$

MV
8860
8862
8863

Fig. 9-6   Connection to a telephone line   (PL)

MV8865

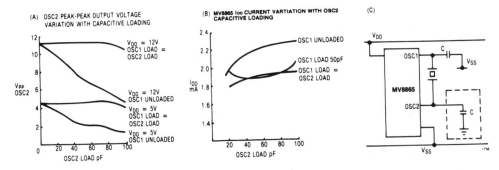

Crystal oscillator loading

Fig. 9-7   (PL)

SL8204

Telephone tone ringer

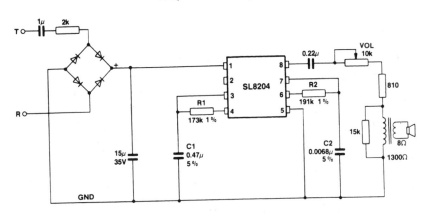

ADDENDUM
ADD 29V ZENER AS
SHOWN FOR HIGH LINE
VOLTAGE PROTECTION

Circuit diagram - tone ringer

Fig. 9-8   (PL)

SL8204

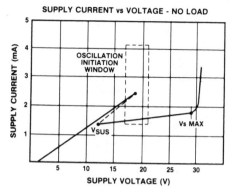

Tone ringer characteristics

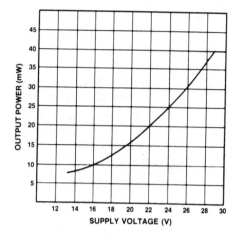

Typical power output

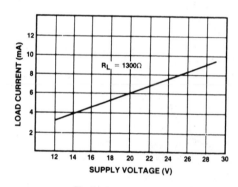

Typical rms current

Fig. 9-9

(PL)

MV4325 KEYPAD PULSE DIALER

The MV4325 Keypad Pulse Dialer contains all the logic necessary to interface a 2 of 7 keypad and convert this key information to control and mute pulses simulating a telephone rotary dial. The MV4325 has programmable access pause capability to provide automatic interruption of dialing needed when accessing the toll network, WATS line or public network. The device is fabricated using Plessey Semiconductors' ISO-CMOS technology which enables the device to function down to 2.0 V making it ideal for long loop operation.

The MV4325 will accept up to 20 digits and access pauses and will redial stored information at a later time by activation of # key. Device current in standby is less than 1 μA at 1.0 V.

The MV4325 is available in Ceramic DIL (DG, $-40°$ C to $+85°$ C).

Applications

☐ Pushbutton Telephones with Last Number Redial
☐ Repertory Dialers
☐ Tone to Pulse Converters

Features

☐ Last Number Redial
☐ Multiple Access Pause Programming
☐ Any Valid Keypad Input or HOLD IN Causes Exit from Access Pause
☐ Oscillator Start Up Controlled from Keypad Input
☐ Oscillator Power Down whilst not Dialing
☐ 300 Hz Key Tone indicates Valid Key
☐ 2.0 V to 7.0 V Supply Voltage Operating Range
☐ Stores up to 20 Digits and Access Pauses
☐ Digit Memory Retained down to 1.0 V at 1 μA
☐ Selectable Mark/Space Ratio 66⅔ : 33⅓ or 60 : 40
☐ 10 Hz Dialing Speed (932 Hz Fast Test):)

MV4325

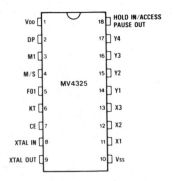

DG18

Fig. 9-10   Pin connections (top view)   (PL)

MV4325

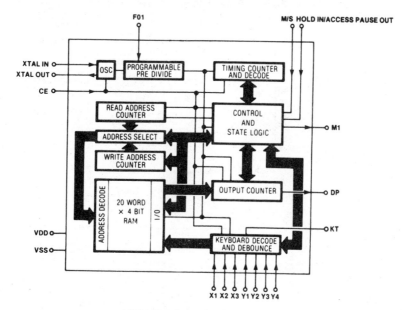

Fig. 9-11

MV4325 function diagram

(PL)

MV4325

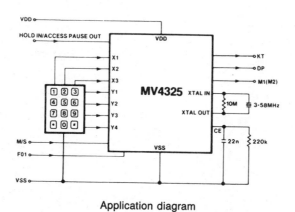

Application diagram

Fig. 9-12

(PL)

XTR100   Precision, low drift 4 mA to 20 mA two-wire transmitter.

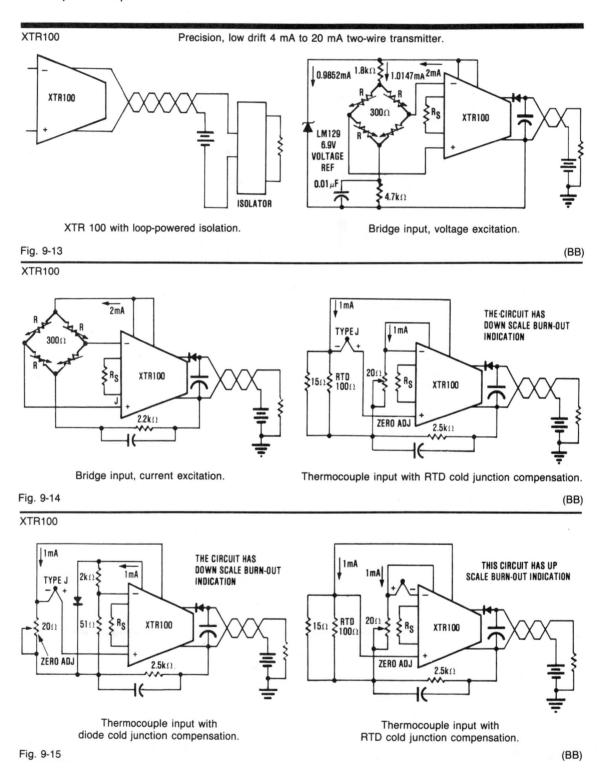

XTR 100 with loop-powered isolation.

Bridge input, voltage excitation.

Fig. 9-13   (BB)

XTR100

Bridge input, current excitation.

Thermocouple input with RTD cold junction compensation.

Fig. 9-14   (BB)

XTR100

Thermocouple input with
diode cold junction compensation.

Thermocouple input with
RTD cold junction compensation.

Fig. 9-15   (BB)

XTR100   Precision, low drift 4 mA to 20 mA two-wire transmitter

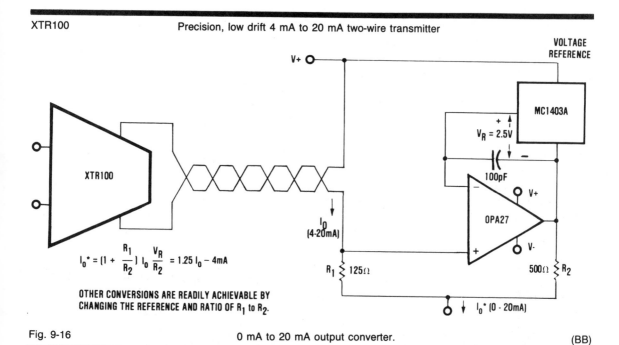

$$I_0{}^* = (1 + \frac{R_1}{R_2})\, I_0\, \frac{V_R}{R_2} = 1.25\, I_0 - 4mA$$

OTHER CONVERSIONS ARE READILY ACHIEVABLE BY CHANGING THE REFERENCE AND RATIO OF R_1 TO R_2.

Fig. 9-16   0 mA to 20 mA output converter.   (BB)

2-WIRE CURRENT OUTPUT TEMPERATURE TRANSDUCER

The AD590 is an integrated-circuit temperature transducer which produces an output current proportional to absolute temperature. The device acts as a high impedance constant current regulator, passing 1 μA/° K for supply voltages between +4 V and +30 V. Laser trimming of the chip's thin film resistors is used to calibrate the device to 298.2 μA output at 298.2° K (+25° C).

The AD590 should be used in any temperature-sensing application between −55° C and +150° C (0° C and 70° C for TO-92) in which conventional electrical temperature sensors are currently employed. The inherent low cost of a monolithic integrated circuit combined with the elimination of support circuitry makes the AD590 an attractive alternative for many temperature measurement situations. Linearization circuitry, precision voltage amplifiers, resistance-measuring circuitry and cold-junction compensation are not needed in applying the AD590. In the simplest application, a resistor, a power source and any voltmeter can be used to measure temperature.

In addition to temperature measurement, applications include temperature compensation or correction of discrete components, and biasing proportional to absolute temperature. The AD590 is available in chip form making it suitable for hybrid circuits and fast temperature measurements in protected environments.

The AD590 is particularly useful in remote sensing applications. The device is insensitive to voltage drops over long lines due to its high-impedance current output. Any well-insulated twisted pair is sufficient for operation hundreds of feet from the receiving circuitry. The output characteristics also make the AD590 easy to multiplex: the current can be switched by a CMOS multiplexer or the supply voltage can be switched by a logic gate output.

AD590

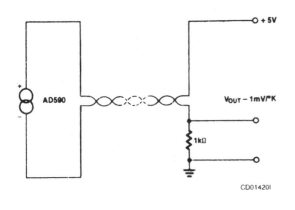 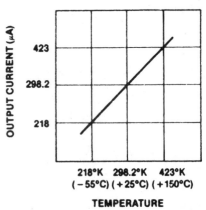

CD01420I

OP01620I

Simple connection. Output is proportional to absolute temperature.

Fig. 9-17

AD590

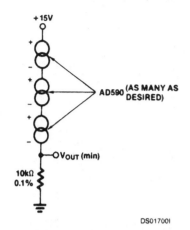

DS01700I

Lowest-temperature sensing scheme.
Available current is that of the "coldest" sensor.

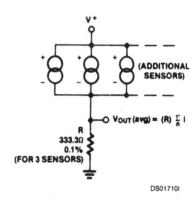

DS01710I

Average-temperature sensing scheme. The sum of the AD590 currents appears across R, which is chosen by the formula:

$$R = \frac{10 \text{ k}\Omega}{n}$$

n being the number of sensors.

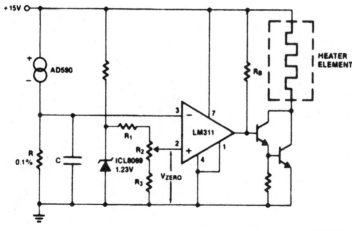

BD00570I

Single-setpoint temperature controller. The AD590 produces a temperature-dependent voltage across R (C is for filtering noise). Setting R2 produces a scale-zero voltage. For the Celsius scale, make R = 1 kΩ and V_{ZERO} = 0.273 volts. For Fahrenheit, R = 1.8 kΩ and V_{ZERO} = 0.460 volts.

Fig. 9-18

AD590

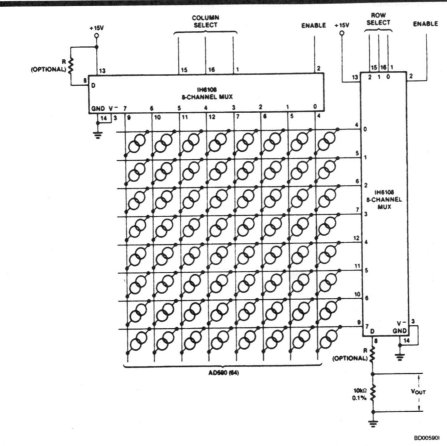

Multiplexing sensors. If shorted sensors are possible, a series resistor in series with the D line will limit the current (shown as R, above: only one is needed). A six-bit digital word will select one of 64 sensors.

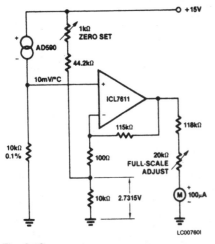

Centigrade thermometer (0° C-100° C). The ultra-low bias current of the ICL7611 allows the use of large-value gain-resistors, keeping meter-current error under ½%, and therefore saving the expense of an extra meter-driving amplifier.

Fig. 9-19

AD590

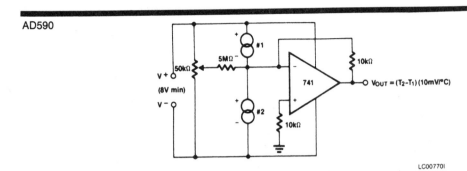

LC007701

Differential thermometer. The 50 kΩ pot trims offsets in the devices whether internal or external, so it can be used to set the size of the difference interval. This also makes it useful for liquid-level detection (where there will be a measurable temperature difference).

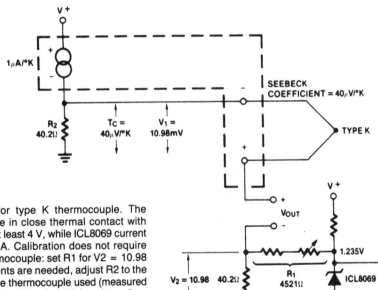

Cold-junction compensation for type K thermocouple. The reference junction(s) should be in close thermal contact with the AD590 case. V + must be at least 4 V, while ICL8069 current should bet set at 1 mA – 2 mA. Calibration does not require shorting or removal of the thermocouple: set R1 for V2 = 10.98 mV. If very precise measurements are needed, adjust R2 to the exact Seebeck coefficient for the thermocouple used (measured or from table) note V1, and set R1 to buck out this voltage (i.e., set V2 = V1). For other thermocouple types, adjust values to the appropriate Seebeck coefficient.

DS017201

LC007801

Simplest thermometer. Metere displays current output directly in degrees Kelvin. Using the AD590M, sensor output is within ± 1.7 degrees over the entire range, and less than ± 1 degree over the greater part of it.

Fig. 9-20

(IN)

AD590

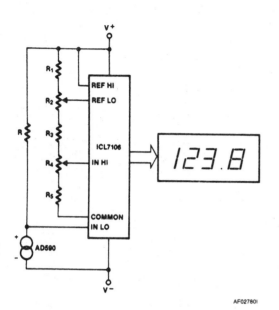

	R	R₁	R₂	R₃	R₄	R₅
°F	9.00	4.02	2.0	12.4	10.0	0
°C	5.00	4.02	2.0	5.11	5.0	11.8

$$\sum_{n=1}^{5} R_n = 28k\Omega \text{ nominal}$$

All values in kΩ

The ICL7106 has a V_{IN} span of ±2.0V, and a V_{CM} range of $(V^+ -0.5)$ Volts to $(V^- +1)$ Volts; R is scaled to bring each range within V_{CM} while not exceeding V_{IN}. V_{REF} for both scales is 500mV. Maximum reading on the Celsius range is 199.9°C, limited by the (short-term) maximum allowable sensor temperature. Maximum reading on the Fahrenheit range is 199.9°F (93.3°C), limited by the number of display digits. See note next page.

AF027801

Basic digital thermometer, Celsius and Fahrenheit scales

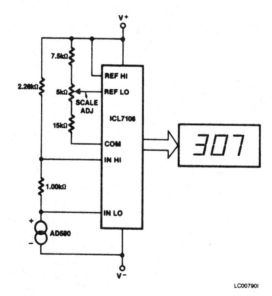

LC007901

Basic digital thermometer, Kelvin scale. The Kelvin scale version reads from 0 to 1999° K theoretically, and from 223° K to 473° K actually. The 2.26 kΩ resistor brings the input within the ICL7106 V_{CM} range: 2 general-purpose silicon diodes or an LED may be substituted.

Fig. 9-21

AD590

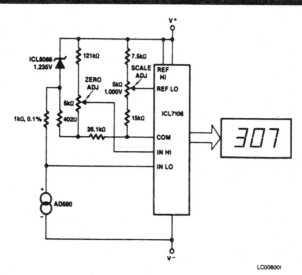

LC008001

Basic digital thermometer, Kelvin scale with zero adjust. This circuit allows "zero adjustment" as well as slope adjustment. The ICL8069 brings the input within the common-mode range, while the 5 kΩ pots trim any offset at 218° K (MS55° C), and set the scale factor.

SCALE	V_{IN} RANGE (V)	R_{INT}(kΩ)	C_{AZ}(μF)
K	0.223 to 0.473	220	0.47
C	−0.25 to +1.0	220	0.1
F	−0.29 to +0.996	220	0.1

For all:

$C_{REF} = 0.1 \mu F$
$C_{INT} = 0.22 \mu F$

$C_{OSC} = 100pF$
$R_{OSC} = 100k\Omega$

Since all 3 scales have narrow V_{IN} spans, some optimization of ICL7106 components can be made to lower noise and preserve CMR. The table below shows the suggested values. Similar scaling can be used with the ICL7126/36.

Fig. 9-22

(IN)

DMOS IN TELEPHONE HANDSET

A telephone handset encounters wide variations of voltage during normal operation. While the dc voltage appearing across the unit may vary from approx. 3 to 25 volts when the phone is off the hook, high voltage ac ringer signals and associated transients have to be handled safely. Moreover, atmospheric disturbances (e.g., lightening) and rf radiations are picked up by the lines inducing high voltages which are suppressed by MOVs, gas discharge tubes, etc. (Not shown in schematic.)

LD0104CNC, a constant current source, limits/stabilizes current to the dialer IC. Low threshold TN05 devices used for the pulser and mute switch operate satisfactorily even at 3 volts. TN0524N3's $I_{D(ON)}$ min. = 100 mA @ V_{GS} = 3 volt is more than adequate for this purpose.

LD0104CNC, TN0524N3

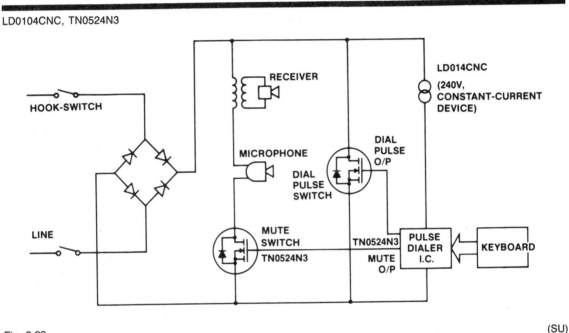

Fig. 9-23

(SU)

ARTIFICIAL NERVE CELL

It's been said much of our technology (sensor logic techniques) strongly resemble nature's own. This circuit of an artificial nerve cell envisions applications in robotics, prosthetic devices, and fast "sense-inference-logic" processors needed for android optical and skin sensors. Sensor or other nerve pulses (or dc) are summed in a non-linear fashion, at the input in coincidence with a threshold pulse, that determines the level of stimulus at which an output of a certain amplitude appears. In addition, supply voltage directly governs circuit action, and is another variable to "nerve" response. A particularly interesting ability of this nerve circuit ,is to time-differentiate the many input pulses by using the threshold pulse in a time-domain function along with its voltage variation. This circuit is useful for lifelike nerve response studies. Supertex DMOS quad arrays make this circuit easy to fabricate with multiple devices in one package.

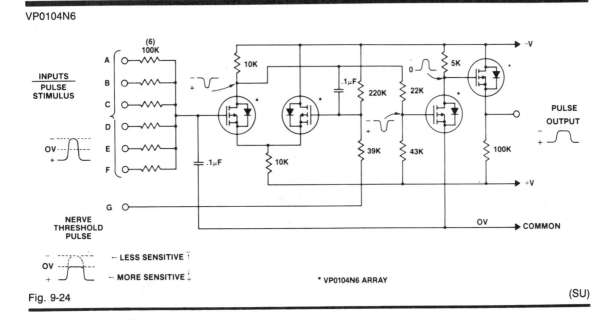

Fig. 9-24   * VP0104N6 ARRAY   (SU)

OPA2111   **ELECTRONIC PROPORTIONAL CONTROLER CIRCUIT**

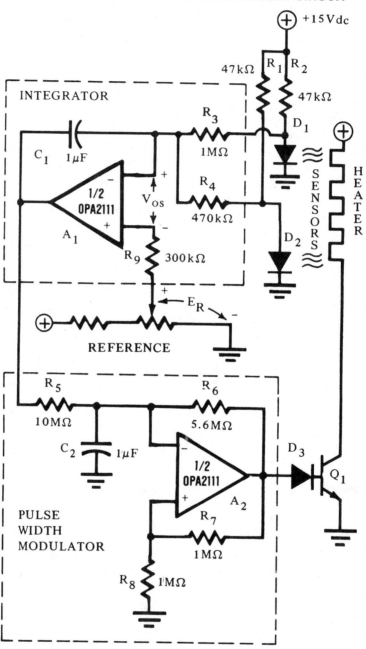

An integrating error amplifier in this proportional controller holds the heater current pulse width at a sustaining value at equilibrium.

Fig. 9-25

(BB)

SL9009

ADAPTIVE CANCELLATION FILTER FOR DUPLEX MODEMS

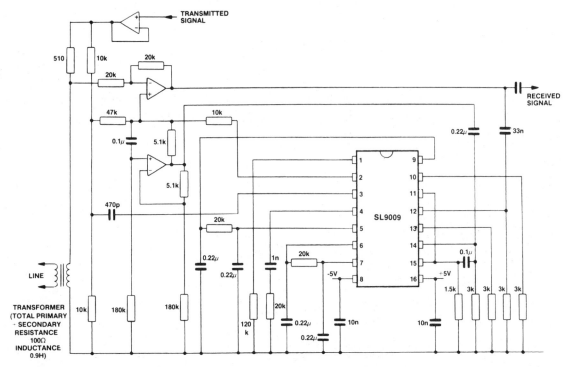

Typical line circuit for modem (900-3000 Hz transmit frequency)

Fig. 9-26

(PL)

V21 MODEM TX/RX

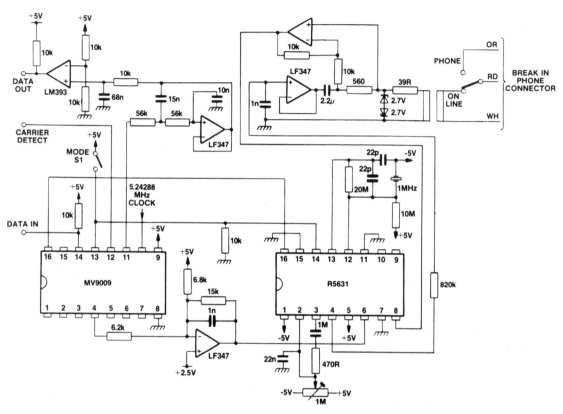

Basic V21 modem

Fig. 9-27

(PL)

MV4320 KEYPAD PULSE DIALER

The MV4320 series is fabricated using ISO-CMOS high density technology. The device is a pin-for-pin replacement for the DF320 Loop Disconnect Dialer and offers wider operating supply voltage range and lower power dissipation. The MV4320 accepts up to 20 digits from a standard 2 of 7 keypad and offers a REDIAL option activated by key #. The device provides dial pulsing and muting outputs and has a HOLD pin for interrupting a dialing sequence. Outpulsing marks/space ratio and dialing speed are pin selectable.

The MV4320 is available in Ceramic DIL (DG, −40° C to +85° C).

Applications

☐ Pushbutton Telephones
☐ Tone to Pulse Converters
☐ Mobile Telephone
☐ Repertory Dialers

Features

☐ Pin for Pin Replacement for the DF320
☐ 2.5 V to 5.5 V Supply Voltage Operating Range
☐ 375 μW Dynamic Power Dissipation at 3 V
☐ Uses Inexpensive 3.58 MHz Ceramic Resonator or Crystal
☐ Stores up to 20 Digits
☐ Selectable Outpulsing Mark/Space Ratio
☐ Selectable Dialing Speeds of 10, 16, 20 and 932 Hz
☐ Low Cost

Operating Notes

The first key entered in any dialing sequence initiates the oscillator by internally taking CE high. Digits may be entered asynchronously from the keypad. Dialing and mute functions are output as shown in Figs. 9-32 and 9-33.

Figure 9-32 shows use of the circuits with external control of CE. This mode is useful if a bistable latching relay is used to mute and switch the complete pulse dialer circuit. In this mode, the pulse occurring on M1 when CE is taken high, with no keypad input, can be used to initiate the bistable latching relay. Figure 9-33 shows the timing diagram for the CE internal control mode. Initially CE is low and goes high on recognition of the first valid key input. Keypad data is entered asynchronously.

Absolute Maximum Ratings

The absolute maximum ratings are limiting values above which operating life may be shortened or specified parameters may be degraded.

MV4320

MV4320

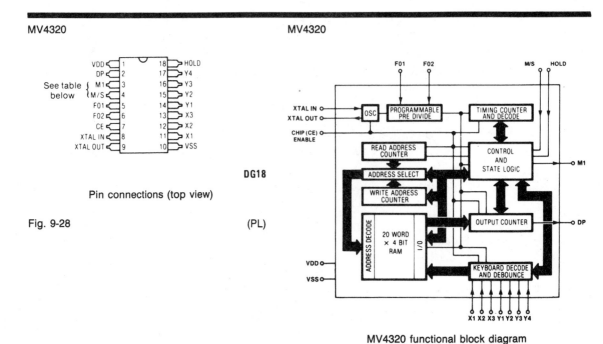

Pin connections (top view)

DG18

Fig. 9-28

(PL)

MV4320 functional block diagram

Fig. 9-29

(PL)

MV4320

		CHARACTERISTICS		SYMBOL	MIN	TYP*	MAX	UNITS	TEST CONDITIONS	
1	S U P P L Y	Supply Voltage Operating Range		V_{DD}	2.5		5.5	V		
2		Standby Supply Current		I_{DDS}		1.0	10.0	µA	$CE = V_{SS}$	
3		Operating Supply Current		I_{DD}		125	200	µA	3.579545 MHz Crystal, $C_{XTALOUT} = 12pF$	
4	I N P U T	Pull-Up Transistor Source Current		I_{IL}	-0.5	-3.0	-12.0	µA	$V_{IN} = V_{SS}$	X_1, X_2, X_3
5		Input Leakage Current		I_{IH}		0.1		nA	$V_{IN} = V_{DD}$	Y_1, Y_2, Y_3, Y_4
6		Input Leakage Current		I_{IL}		-0.1		nA	$V_{IN} = V_{SS}$	M/S, IDP, F01,
7		Pull-Down Transistor Sink Current		I_{IH}	0.5	3.0	12.0	µA	$V_{IN} = V_{DD}$	F02, FD, HOLD
8		Logic '0' Level		V_{iL}			0.9	V	All inputs	
9		Logic '1' Level		V_{IH}	2.1			V		
10	O U T P U T	Voltage Levels	Low-Level	V_{OL}		0	0.01	V	No Load	
11			High-level	V_{OH}	2.99	3		V		
12		Drive Current	N-Channel Sink	I_{OL}	0.8	2.0		mA	$V_{OUT} = 2.3V$	DP, M1/M2
13			P-Channel Source	I_{OH}	-0.8	-2.0		mA	$V_{OUT} = 0.7V$	

Fig. 9-30

(PL)

MV4320

AC ELECTRICAL CHARACTERISTICS

Test conditions (unless otherwise stated):
$V_{DD} = 3.0V$; $T_{amb} = +25°C$; $f_{CLK} = 3.579545\,MHz$
All voltages wrt V_{SS}

		CHARACTERISTICS	SYMBOL	MIN	TYP*	MAX	UNITS	TEST CONDITIONS
14		Output Rise Time	t_R		1.0		us	DP,M$_1$.
15		Output Fall Time	t_F		1.0		us	$C_L = 50pF$
16		Maximum Clock Frequency	t_{CLK}	3.58			MHz	3.579545 MHz Crystal
17	D	Mark to Space Ratio	M/S		2:1			Note 1
18	Y				3:2			
19	N				10			
20	A	Impulsing Rate $= \frac{1}{T}$			16		Hz	Note 1
21	M				20			
22	I				932			
23	C	Clock Start Up Time	t_{on}		1.5	4	ms	Timed from CE '1'
24		Input Capacitance	C_{in}		5.0		pF	Any Input

* Typical parametric values are for Design Aid Only, not guaranteed and not subject to production
testing. Timing waveforms are subject to production functional test.
NOTES:
1. See Pin Function, Table 1.

Fig. 9-31

(PL)

MV4320

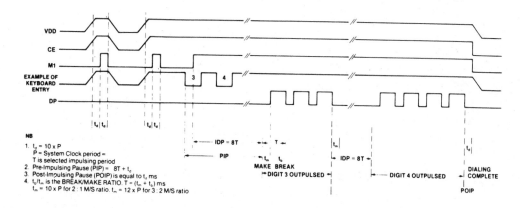

Keypad pulse dialer timing diagram, CE-External control

Fig. 9-32

(PL)

MV4320

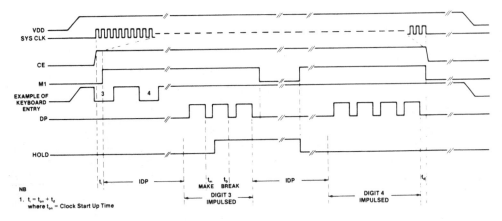

Keypad pulse dialer timing diagram, CE-Internal control

Fig. 9-33 (PL)

MV4320

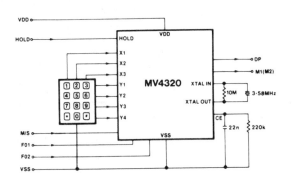

	MIN.	MAX.
V_{DD}-V_{SS}	-0.3V	10V
Voltage on any pin	V_{SS} -0.3V	V_{DD} +0.3V
Current at any pin		10mA
Operating Temperature	-40°C	+85°C
Storage Temperature	-65°C	+150°C
Power Dissipation		1000mW

Derate 16mW/°C above 75°C. All leads soldered to PC board.

Application diagram

Fig. 9-34 (PL)

MV4320

PIN FUNCTIONS

V_{DD}	Positive voltage supply						
DP	Dial Pulsing Output Buffer						
M1	Mute Output (Off Normal) Buffer						
M/S	Mark/Space (Break/Make) Ratio select. On-chip pull-down transistor to V_{SS}.					O/C	2:1
	Note: O/C = Open Circuit					V_{DD}	3:2
F01,F02	Impulsing Rate Selection. On-chip pull-down transistor to V_{SS}. * Assumes f_{CLK} = 3.579545MHz	F01	F02	Nominal Impulsing Rate	Actual* Inpulsing Rate	System Clock frequency	
		O/C	O/C	10Hz	10.13Hz	303.9Hz	
		O/C	V_{DD}	20Hz	19.42Hz	582.6Hz	
		V_{DD}	O/C	932Hz	932.17Hz	27,965.1Hz	
		V_{DD}	V_{DD}	16Hz	15.54Hz	466.1Hz	
CE	Chip Enable. An active input. Control is internal via static keyboard decode, or by external forcing.						
XTAL IN	Crystal Input. Active, clamped low if CE = '0', high impedance if CE = '1'						
XTAL OUT	Crystal Output Buffer to drive crystal.						
V_{SS}	System ground						
X_1,X_2,X_3	Column keyboard Inputs. On-chip pull-up transistors to V_{DD}. Active LOW						
Y_1,Y_2,Y_3,Y_4	Row keyboard Inputs. On-chip pull-up transistors to V_{DD}. Active LOW						
HOLD		O/C	Normal Operation				
		V_{DD}	No impulsing. If activated during impulsing, hold occurs when the current digit is complete				
	Prevents further impulsing. On-chip pull-down transistor to V_{SS}						

Fig. 9-35

(PL)

10
Miscellaneous Circuits

AUTOMATIC BATTERY BACK-UP SWITCH

The Intersil ICL7673 is a monolithic CMOS battery backup circuit that offers unique performance advantages over conventional means of switching to a backup supply. The ICL7673 is intended as a low-cost solution for the switching of systems between two power supplies; main and battery backup.

The main application is keep-alive-battery power switching for use in volatile CMOS RAM memory systems and real time clocks. In many applications this circuit will represent a low insertion voltage loss between the supplies and load.

This circuit features low current consumption, wide operating voltage range, and exceptionally low leakage between inputs. Logic outputs are provided that can be used to indicate which supply is connected and can also be used to increase the power switching capability of the circuit by driving external pnp transistors.

The ICL7673 is available in either an 8-pin plastic minidip package, a TO-99 metal can, or as dice.

Features

☐ Automatically Connects Output to The Greater of Either Input Supply Voltage
☐ If Main Power to External Equipment is Lost, Circuit Will Automatically Connect Battery Backup
☐ Reconnects Main Power When Restored
☐ Logic Indicator Signaling Status Of Main Power
☐ Low Impedance Connection Switches
☐ Low Internal Power Consumption
☐ Wide Supply Range: 2.5 to 15 Volts
☐ Low Leakage Between Inputs
☐ External Transistors May Be Added If Very Large currents Need to Be Switched

Applications

☐ On Board Battery Backup for Real-Time Clocks, Timers, or Volatile RAMs
☐ Over/Under Voltage Detector
☐ Peak Voltage Detector
☐ Other Uses:
—Portable Instruments, Portable Telephones, Line Operated Equipment

Detailed Description

As shown in the functional diagram (Fig. 10-1), the ICL7673 includes a comparator which senses the input voltages V_P and V_S. The output of the comparator drives the first inverter and the open-drain n-channel transistor P_{bar}. The first inverter drives a large p-channel switch, P1, a second inverter, and another open-drain n-channel transistor, S_{bar}.

The second inverter drives another large p-channel switch P2. The ICL7673, connected to a main and a backup power supply, will connect the supply of greater potential to its output. The circuit provides break-before-make switch action as it switches from main to backup power in the event of a main power supply failure.

For proper operation, inputs V_P and V_S must not be allowed to float, and, the difference in the two supplies must be greater than 50 millivolts. The leakage current through the reverse biased parasitic diode of switch P2 is very low.

Output Voltage

The output operating voltage range is 2.5 to 15 volts. The insertion loss between either input and the output is a function of load current, input voltage, and temperature. This is due to the p-channels being operated in their triode region, and, the ON-resistance of the switches is a function of output voltage V_o. The ON-resistance of the p-channels have positive temperature coefficients, and therefore as temperature increases the insertion loss also increases. At low load currents the output voltage is nearly equal to the greater of the two inputs.

The maximum voltage drop across switch P1 or P2 is 0.5 volts, since above this voltage the body-drain parasitic diode will become forward biased. Complete switching of the inputs and open-drain outputs typically occurs in 50 microseconds.

Input Voltage

The input operating voltage range for V_P or V_S is 2.5 to 15 volts. The input supply voltage (V_P or V_S) slew rate should be limited to 2 volts per microsecond to avoid potential harm to the circuit. In line-operated systems, the rate-of-rise (or fall) of the supply is a function of power supply design.

For battery applications it may be necessary to use a capacitor between the input and ground pins to limit the rate-of-rise of the supply voltage. A low-impedance capacitor such as a 0.047 μF disc ceramic can be used to reduce the rate-of-rise.

Status Indicator Outputs

The n-channel open drain output transistors can be used to indicate which supply is connected, or can be used to drive external pnp transistors to increase

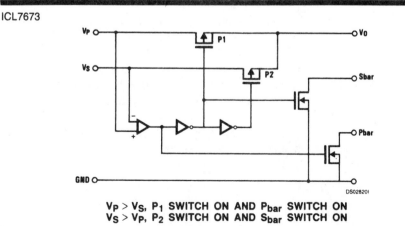

ICL7673

$V_P > V_S$, P_1 SWITCH ON AND P_{bar} SWITCH ON
$V_S > V_P$, P_2 SWITCH ON AND S_{bar} SWITCH ON

Fig. 10-1   Functional diagram   (IN)

the power switching capability of the circuit. When using external pnp power transistors, the output current is limited by the beta and thermal characteristics of the power transistors. The application section details the use of external pnp transistors.

Applications

A typical discrete battery backup circuit is illustrated in Fig. 10-6. This approach requires several components, substantial printed circuit board space, and high labor cost. It also consumes a fairly high quiescent current.

The ICL7673 battery backup circuit, illustrated in Fig. 10-5, will often replace such discrete designs and offer much better performance, higher reliability, and lower system manufacturing cost.

A trickle charge system could be implemented with an additional resistor and diode as shown in Fig. 10-7. A complete low power ac to regulated dc system can be implemented using the ICL7673 and ICL7663 micropower voltage regulator as shown in Fig. 10-8.

Applications for the ICL7673 include volatile semiconductor memory storage systems, real-time clocks, timers, alarm systems, and over/under voltage detectors. Other systems requiring dc power when the master ac line supply fails can also use the ICL7673.

A typical application, as illustrated in Fig. 10-9, would be a microprocessor system requiring a 5 volt supply. In the event of primary supply failure, the system is powered down, and a 3 volt battery is employed to maintain clock or volatile memory data.

The main and backup supplies are connected to V_p and V_s, with the circuit output V_o supplying power to the clock or volatile memory. The ICL7673 will sense the main supply, when energized, to be of greater potential than V_s and connect, via its internal MOS switches, V_p to output V_o.

The backup input, V_s will be disconnected internally. In the event of main supply failure, the circuit will sense that the backup supply is now the greater potential, disconnect V_p from V_o, and connect V_s.

Figure 10-10 illustrates the use of external pnp power transistors to increase the power switching capability of the circuit. In this application the output current is limited by the beta and thermal characteristics of the power transistors.

If hysteresis is desired for a particular low power application, positive feedback can be applied between the input V_p and open drain output S_{bar} through a resistor as illustrated in Fig. 10-11. For high power applications hysteresis can be applied as shown in Fig. 10-12.

The ICL7673 can also be used as a clipping circuit as illustrated in Fig. 10-13. With high impedance loads the circuit output will be nearly equal to the greater of the two input signals.

ICL7673

ABSOLUTE MAXIMUM RATINGS

Input Supply (V_P or V_S) Voltage .... –0.3 to +18V
Output Voltages P_{bar} and S_{bar} .... –0.3 to +18V
Peak Current
    Input V_P (@ V_P = 5V) (note 1) 38mA
    Input V_S (@ V_S = 3V) 30mA
    P_{bar} or S_{bar} 150mA
Continuous Current
    Input V_P (@ V_P = 5V) (note 1) 38mA
    Input V_S (@ V_S = 3V) 30mA
    P_{bar} or S_{bar} 50mA

Package Dissipation 300mW
    Linear Derating Factors
    TO-99   PLASTIC
    5.7mW/°C   6.1mW/°C
    above 50°C   above 36°C
Operating Temperature Range:
    ICL7673CPA/CBA .... 0°C to +70°C
    ICL7673ITV .... –25°C to +85°C
Storage Temperature .... –65°C to +150°C
Lead Temperature (Soldering, 10sec) .... 300°C
Note 1. Derate above 25°C by 0.38mA/°C.

Stresses above those listed under "Absolute Maximum Ratings" may cause permanent damage to the device. These are stress ratings only and functional operation of the device at these or any other conditions above those indicated in the operational sections of the specifications is not implied. Exposure to absolute maximum rating conditions for extended periods may affect device reliability.

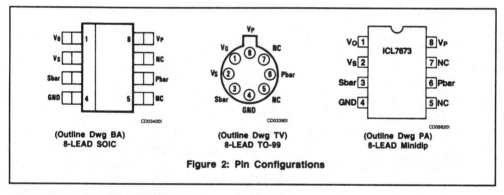

Figure 2: Pin Configurations

ELECTRICAL CHARACTERISTICS (T_A = 25°C unless otherwise specified)

SYMBOL	PARAMETER	TEST CONDITIONS	MIN	TYP	MAX	UNIT
V_P	INPUT VOLTAGE	V_S = 0 volts I load = 0mA	2.5	–	15	V
V_S		V_P = 0 volts I load = 0mA	2.5	–	15	
I^+	QUIESCENT SUPPLY CURRENT	V_P = 0 volts V_S = 3 volts I load = 0mA	–	1.5	5	µA
$R_{ds(on)}P_1$	SWITCH RESISTANCE P1 (NOTE 2)	V_P = 5 volts V_S = 3 volts I load = 15mA	–	8	15	Ω
		@ T_A = 85°C	–	16	–	
		V_P = 9 volts V_S = 3 volts I load = 15mA	–	6	–	Ω
		V_P = 12 volts V_S = 3 volts I load = 15mA	–	5	–	Ω
$T_{C(P1)}$	TEMPERATURE COEFFICIENT OF SWITCH RESISTANCE P1	V_P = 5 volts V_S = 3 volts I load = 15mA	–	2.03	–	%/°C

Pin configurations

Fig. 10-2

(IN)

ICL7673

ELECTRICAL CHARACTERISTICS (CONT.)

SYMBOL	PARAMETER	TEST CONDITIONS	MIN	TYP	MAX	UNIT
$R_{ds(on)P2}$	SWITCH RESISTANCE P2 (NOTE 2)	$V_P = 0$ volts $V_S = 3$ volts I load = 1mA	–	40	100	Ω
		@ $T_A = 85°C$	–	60	–	
		$V_P = 0$ volts $V_S = 5$ volts I load = 1mA	–	26	–	Ω
		$V_P = 0$ volts $V_S = 9$ volts I load = 1mA	–	16	–	Ω
$T_{C(P2)}$	TEMPERATURE COEFFICIENT OF SWITCH RESISTANCE P2	$V_P = 0$ volts $V_S = 3$ volts I load = 1mA	–	0.7	–	%/°C
$I_{L(PS)}$	LEAKAGE CURRENT (V_P to V_S)	$V_P = 5$ volts $V_S = 3$ volts I load = 10mA	–	0.01	20	nA
		@ $T_A = 85°C$	–	35	–	
$I_{L(SP)}$	LEAKAGE CURRENT (V_S to V_P)	$V_P = 0$ volts $V_S = 3$ volts I load = 1mA	–	0.01	50	nA
		@ $T_A = 85°C$	–	120	–	
$V_{O\,Pbar}$	OPEN DRAIN OUTPUT SATURATION VOLTAGES	$V_P = 5$ volts $V_S = 3$ volts I sink = 3.2mA I load = 0mA	–	85	400	mV
		@ $T_A = 85°C$	–	120	–	
		$V_P = 9$ volts $V_S = 3$ volts I sink = 3.2mA I load = 0mA	–	50	–	mV
		$V_P = 12$ volts $V_S = 3$ volts I sink = 3.2mA I load = 0mA	–	40	–	mV
$V_{O\,Sbar}$		$V_P = 0$ volts $V_S = 3$ volts I sink = 3.2mA I load = 0mA	–	150	400	mV
		@ $T_A = 85°C$	–	210	–	
		$V_P = 0$ volts $V_S = 5$ volts I sink = 3.2mA I load = 0mA	–	85	–	mV
		$V_P = 0$ volts $V_S = 9$ volts I sink = 3.2mA I load = 0mA	–	50	–	mV

Fig. 10-3

ICL7673

ELECTRICAL CHARACTERISTICS (CONT.)

SYMBOL	PARAMETER	TEST CONDITIONS	MIN	TYP	MAX	UNIT
I_L Pbar	OUTPUT LEAKAGE CURRENTS OF Pbar AND Sbar	V_P = 0 volts V_S = 15 volts I load = 0mA	–	50	500	nA
		@ T_A = 85°C	–	900	–	
I_L Sbar		V_P = 15 volts V_S = 0 volts I load = 0mA	–	50	500	nA
		@ T_A = 85°C		900	–	
$V_P - V_S$	SWITCHOVER UNCERTAINTY FOR COMPLETE SWITCHING OF INPUTS AND OPEN DRAIN OUTPUTS.	V_S = 3 volts I sink = 3.2mA I load = 0mA	–	5	50	mV

NOTE 2. The minimum input to output voltage can be determined by multiplying the load current by the switch resistance.

TYPICAL PERFORMANCE CHARACTERISTICS

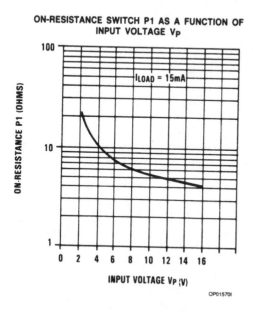

ON-RESISTANCE SWITCH P1 AS A FUNCTION OF INPUT VOLTAGE V_P

OP01570I

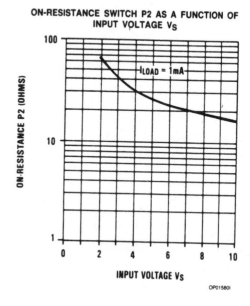

ON-RESISTANCE SWITCH P2 AS A FUNCTION OF INPUT VOLTAGE V_S

OP01580I

Fig. 10-4

ICL7673

ICL7673

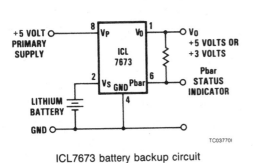

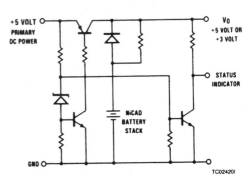

ICL7673 battery backup circuit

Discrete battery backup circuit

Fig. 10-5   (IN)   Fig. 10-6   (IN)

ICL7673

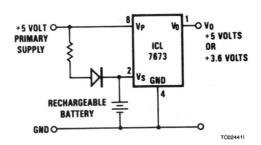

Application requiring rechargeable battery backup

Fig. 10-7   (IN)

ICL7673

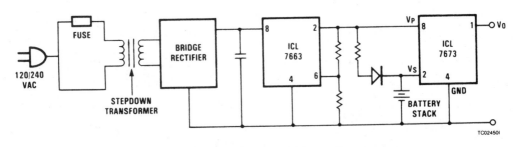

Power supply for low power portable ac to dc systems

Fig. 10-8   (IN)

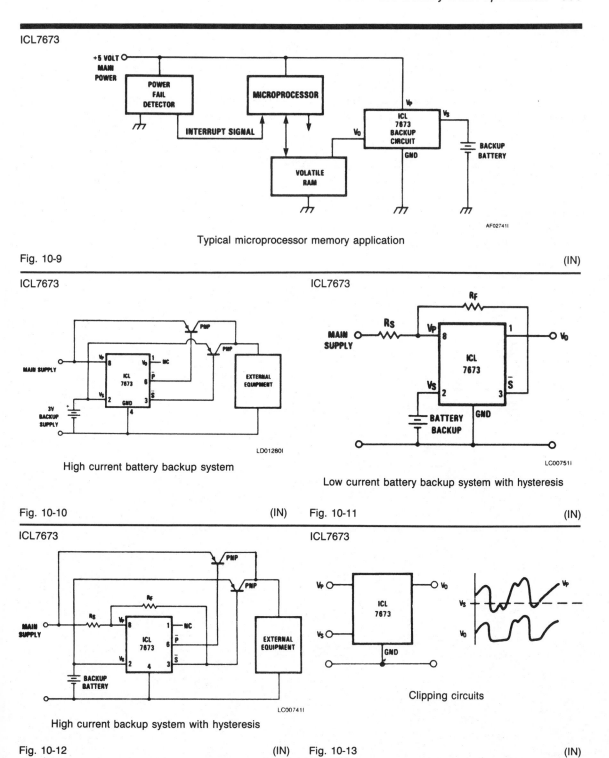

ICL7673

Typical microprocessor memory application

Fig. 10-9 (IN)

ICL7673

High current battery backup system

Fig. 10-10 (IN)

ICL7673

Low current battery backup system with hysteresis

Fig. 10-11 (IN)

ICL7673

High current backup system with hysteresis

Fig. 10-12 (IN)

ICL7673

Clipping circuits

Fig. 10-13 (IN)

SPEED REGULATOR FOR DC MOTORS

The TDA1154 is a monolithic integrated circuit intended for speed regulation of permanent magnet dc motors used in record players, tape recorders, cassette recorders and toys.

The circuit offers an excellent speed regulation with much higher power supply, temperature and load variations than conventional circuits built around discrete components.

- ☐ Matching flexibility to motors with various characteristics.
- ☐ Built-in current limit
- ☐ On-chip 1.2 V reference voltage
- ☐ Starting current: 0.5 A @ 2.5 V
- ☐ Reflection coefficient K = 20
- ☐ Supply voltage: $\leq$ + 20 V

TDA1154

MAXIMUM RATINGS

Rating	Symbol	Value	Unit
Supply voltage	V_{CC}	20	V
Output current	I_O	1.2	A
Power dissipation	P_{tot}	(see curve)	W
Junction temperature	T_j	+ 150	°C
Storage temperature range	T_{stg}	− 55 to + 150	°C

THERMAL CHARACTERISTICS

Characteristic	Symbol	Value	Unit
Junction-ambient thermal resistance	$R_{th(j-a)}$	110	°C/W
Junction-case thermal resistance	$R_{th(j-c)}$	19	°C/W

Fig. 10-14   (TH)

TDA1154

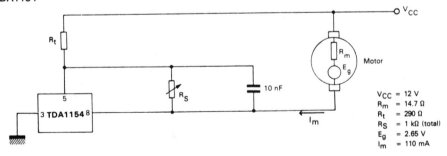

V_{CC} = 12 V
R_m = 14.7 Ω
R_t = 290 Ω
R_S = 1 kΩ (total)
E_g = 2.65 V
I_m = 110 mA

3 : Ground
5 : Reference
8 : Output
Other pins are not connected

Fig. 10-15   (TH)

The circuit maintains a 1.2 V constant reference voltage between pins 5 and 8 :

$$V(5 - 8) = V_{(ref)} = 1.2 \text{ V}$$

The current (I(5)) drawn by the circuit at pin 5 is the sum of two currents. One is constant: $I_0(5) = 1.7$ mA and the other is proportional to pin 8 current (I(8)) :

$$I(5) = I_0(5) + I(8) \text{ K (a)} \quad (I_0(5) = 1.7 \text{ mA, K} = 20)$$

If E_g and R_m are motor back electromotive force and motor internal resistance respectively, then:

$$E_g + R_m I_m = R_t \left[I(5) + \frac{V_{(ref)}}{R_S} \right] + V_{(ref)} \text{ (b)}$$

From Fig. 10-17 it is seen that:

$$I(8) = I_m + \frac{V_{(ref)}}{R_S} \text{ (c)}$$

Substituting equations (a) and (c) into (b) yields:

$$E_g = I_m \underbrace{\left[\frac{R_t}{K} - R_m \right]}_{(1)} + V_{(ref)} \underbrace{\left[\frac{R_t}{R_S} \left(1 + \frac{1}{K} \right) + 1 \right] + R_t I_0(5)}_{(2)} \text{ (d)}$$

The motor speed will be independent of the resisting torque if E_g is also independent of I_m. Therefore, in order to determine the value of R_t term (1) in (d) must be zero:

$$\boxed{R_t = K \ R_m \ (K = 20)}$$

If $R_t > K \ R_m$, an instability may occur as a result of overcompensation.

The value of R_S is determined by term (2) in (d) so as to obtain the back electromotive force (E_g) corresponding to required motor speed:

$$R_S = R_t \ \frac{V_{(ref)} (1 + 1/K)}{E_g - V_{(ref)} - R_t I_0(5)} \neq R_t \ \frac{V_{(ref)}}{E_g - V_{(ref)} - R_t I_0(5)}$$

where $V_{(ref)} = 1.2$ V and $I_0(5) = 1.7$ mA

TDA1154

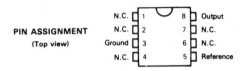

PIN ASSIGNMENT
(Top view)

N.C.	1	8	Output
N.C.	2	7	N.C.
Ground	3	6	N.C.
N.C.	4	5	Reference

Fig. 10-16   (TH)

TDA1154

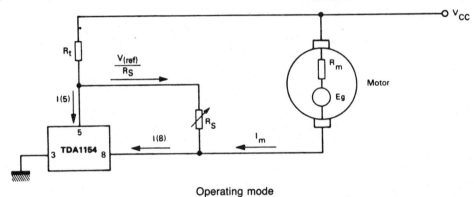

Operating mode

Fig. 10-17   (TH)

TDA1154

ELECTRICAL CHARACTERISTICS

$T_{amb} = +25°C$
(Unless otherwise specified)

Characteristic	Symbol	Min	Typ	Max	Unit
Reference voltage ($V_{CC} = +6$ V, $I(8) = 0.1$ A)	$V_{(ref)}$	1.15	1.25	1.35	V
Reference voltage temperature coefficient $V_{CC} = +6$ V, $I(8) = 0.1$ A, $T_{amb} = -20°C$ to $+70°C$	$\frac{\Delta V_{(ref)}}{V_{(ref)}} / \Delta T$	—	0.02	—	%/°C
Line regulator ($V_{CC} = +4$ V to $+18$ V, $I(8) = 0.1$ A)	$\frac{\Delta V_{(ref)}}{V_{(ref)}} / \Delta V_{CC}$	—	0.02	—	%/°C
Load regulator ($V_{CC} = +6$ V, $I(8) = 25$ to 400 mA)	$\frac{\Delta V_{(ref)}}{V_{(ref)}} / \Delta I(8)$	—	0.009	—	%/mA
Minimum supply voltage ($I(8) = 0.1$ A, $\frac{\Delta V_{(ref)}}{V_{(ref)}} = -5\%$)	$V(5-3)$	2.5	—	—	V
Starting current (*) ($\frac{\Delta V_{(ref)}}{V_{(ref)}} = -50\%$) $V_{CC} = +5$ V	$I(8)$	1.2	—	—	A
$V_{CC} = +2.5$ V	$I(8)$	0.5	0.8	—	A
Quiescent current on pin 5 ($V_{CC} = +6$ V, $I(8) = 100$ μA)	$I_O(5)$	—	1.7	—	mA
$K = \frac{\Delta I(8)}{\Delta I(5)}$ reflection coefficient ($V_{CC} = +6$ V, $I(8) = 0.1$ A)	K	18	20	22	
K spread versus V_{CC} ($V_{CC} = +6$ V to $+18$ V, $I(8) = 0.1$ A)	$\frac{\Delta K}{K} / \Delta V_{CC}$		0.45	—	%/V
K spread versus $I(8)$ ($V_{CC} = +6$ V, $I(8) = 25$ to 400 mA)	$\frac{\Delta K}{K} / \Delta I(8)$	—	0.005	—	%/mA
K spread versus temperature $V_{CC} = +6$ V, $I(8) = 0.1$ A, $T_{amb} = +20°C$ to $+70°C$	$\frac{\Delta K}{K} / \Delta T$	—	0.02	—	%/°C

(*) An internal protection circuit reduces this current if the temperature of the junction increase : $I(8) = 0.75$ A at $T_j = +140°C$

Fig. 10-18   (TH)

TDA1154

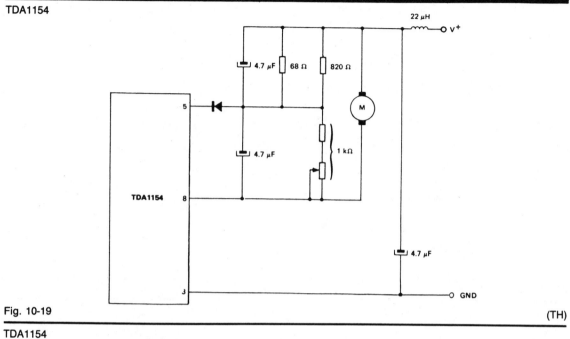

Fig. 10-19 (TH)

TDA1154

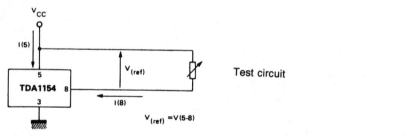

Test circuit

$V_{(ref)} = V(5-8)$

Fig. 10-20 (TH)

TDA1154

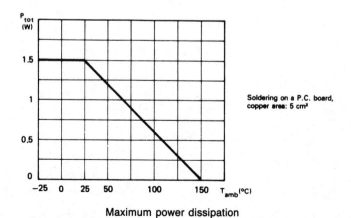

Maximum power dissipation

Fig. 10-21 (TH)

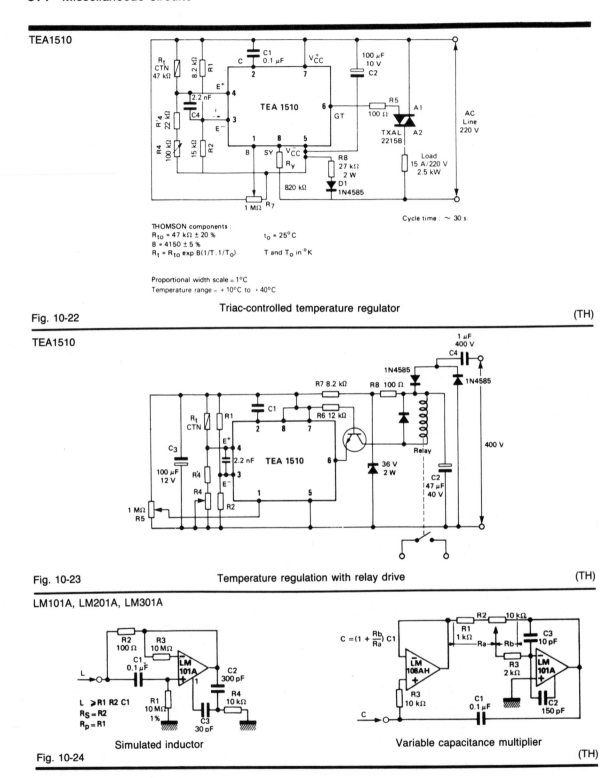

TEA1510

THOMSON components :
R_{to} = 47 kΩ ± 20 % t_o = 25°C
B = 4150 ± 5 %
R_t = R_{to} exp B(1/T . 1/T_o)   T and T_o in °K

Proportional width scale = 1°C
Temperature range = + 10°C to + 40°C

Cycle time : ~ 30 s.

Fig. 10-22   Triac-controlled temperature regulator   (TH)

TEA1510

Fig. 10-23   Temperature regulation with relay drive   (TH)

LM101A, LM201A, LM301A

L ⩾ R1 R2 C1
R_S = R2
R_p = R1

Simulated inductor

$C = (1 + \dfrac{Rb}{Ra})$ C1

Variable capacitance multiplier

Fig. 10-24   (TH)

CURRENT REGULATED FAST PRINT MATRIX

Fast rate printing heads utilizing solenoids experience output variations with temperature. Current regulation during operation greatly improves performance, but increases the circuitry needed for large matrices. In this application, Supertex VN01 array functions as column driver and current regulator, reducing parts significantly while improving response time for this fast printer. Support circuitry driving the 48 rows is also greatly reduced with the AN01 array, containing 8 drivers per package. The VP0220 transistors powering the row connected coils, provide fast switching and reliability, with their 200 volt rating and −85 Amp pulse ability.

With the actual switching scheme TTL controllable, current regulation is controlled by I_D (ON) of the VN0106 and the inherent V_{GS} (th) match. The gating diodes and variable voltage source modify the voltage actually appearing on that gate, therefore controlling maximum conducted drain current. The VN0106 array, rated at 60 volts, requires voltage clamp diodes to limit inductive-voltage peaks appearing on the drains (shown only on pin #14).

Another important feature of this circuit limits the gate voltage applied to the VP02 row drivers (2 transistors shown on schematic). At first glance, it seems the gate source voltage rating is exceeded when the AN01 turns on, grounding the driver's gate-producing 28 volt gate over-drive. In reality, the AN01's 200-300 ohm R_{DS} ON divides the gate drive to a safe 16 V level, eliminating an external resistor function.

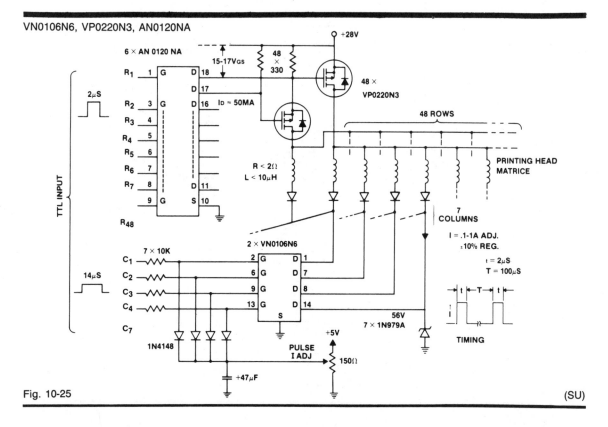

VN0106N6, VP0220N3, AN0120NA

Fig. 10-25

ELECTROCARDIOGRAM PROTECTOR

Electrocardiogram amplifiers must process 1 millivolt signals while withstanding electrosurgery and defibrilator pulses of 1000 to 8000 volts, and currents of several amperes. This circuit provides the necessary common-mode protection with transistors A & B, and differential limiting with C & D. Amplifier protection, with greatly reduced overload recovery time, justifies the increased complexity over conventional spark gaps and diodes.

Protection clamping is initiated when the protect enable circuit detects an overload, powering-on the 2 MHz oscillator, which produces gate drive through a specially designed medical isolation transformer. Ac clamping action is performed by the low $R_{DS(ON)}$ of the VN1220 transistors during forward conduction. Reverse polarity conduction through the DMOS Body Diode is enhanced by the gate drive, producing a V_{SD} of about 10 millivolts for the 100 mA I_{SD}.

The three stage attenuator in this design provides greater than 10,000:1 voltage reduction and input error due to H.V. noise; representing a typical 1 millivolt error signal at the amplifier input. This circuit, whole or in part, is useful for signal processing circuitry exposed to repeated voltage and current overloads.

VN1220N2

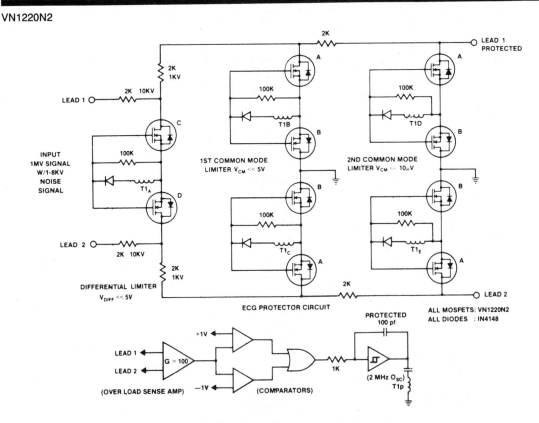

Fig. 10-26   Protect enable circuit   (SU)

50 MICROWATT TEMPERATURE CONTROLLER

This heating/cooling machinery control circuit features only 50 microwatts power consumption and a linear dial function. The ability of this circuit to control several amperes of current and retain battery life is outstanding. Shown are just three of many possible output configurations using p-channel DMOS. $R_{DS(ON)}$ and BV_{DSS} requirements determine proper DMOS device selection.

VP0535N3, VP1106N5, VP1204N5

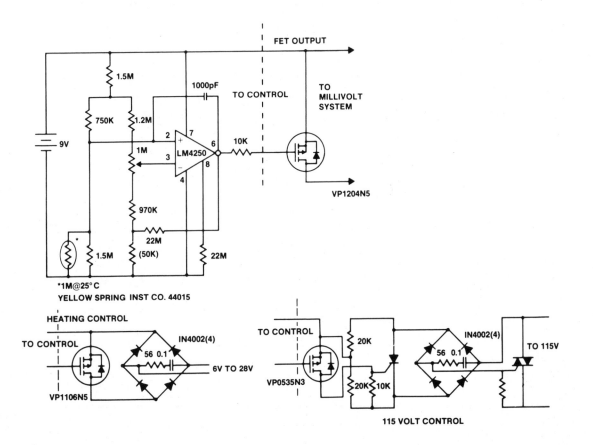

Fig. 10-27

(SU)

IMPROVED RF PULSER

Everyone faced with the generation of fast power-pulse circuitry will be particularly interested in this application. A TTL pulse is inputed to IC 1 and is accurately level-shifted to a −15 volts referenced pulse at collectors of Q2 & Q3. Q4 functions by turning off Q5 and Q6. The schottky diodes provide a low impedance discharge for DMOS gate capacitance through the collector of Q4, and as a Baker clamp technique for Q5.

The combination of DMOS technology and fast driver techniques, allows this circuit to replace an expensive hybrid usually used for this application. Cost savings can be as much as 4 times. The driver section of this circuit is applicable to most DMOS devices, particularly where very-fast rise and fall pulses are needed, such as research instrumentation, high-frequency switching power supplies, rf communication circuits, etc.

VN10KN3

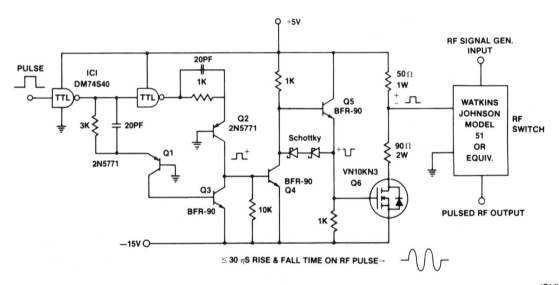

Fig. 10-28   (SU)

RANGEFINDER POWER BOOSTER

The well known LM1812 rangefinder IC is given an order of magnitude more drive current, and better coil damping, using the VP12 as a booster stage. The need for a transmit decoupling capacitor on pin 13 is eliminated, while the body diode of the DMOS transistor quickly damps out ringing potentials of the transmit coil even when they exceed the supply voltage.

A variety of applications include: Sonar, process control, chemical vat/air/gas temperature monitors and ultrasound data links.

VP1204N2

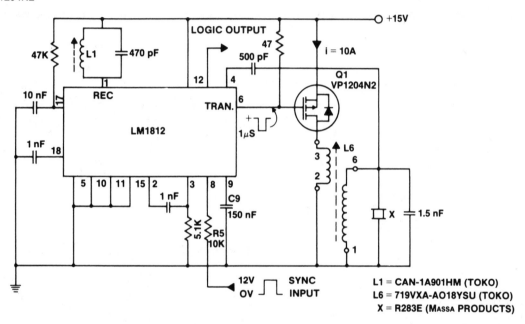

Fig. 10-29 (SU)

50-240 HZ INVERTER WITH PRECISION SNUBBER

Most switching power converters utilizing more than one power transistor need "dead time" before the next device is switched on. This allows the transformer ringing signal to dissipate in appropriate passive components, and not in the switching transistors.

Dead time is digitally generated along with the DMOS drive signals, with U1-4 allowing the inverter's frequency to be varied and keep dead time at 11.25 degrees. U5 is of particular interest, as the 74C908 IC sources over ¼ amp, and along with R3 & R4 provide optimum DMOS switching performance. Higher output power requirements are met by paralleling VN0335N1 devices.

VN0335N1

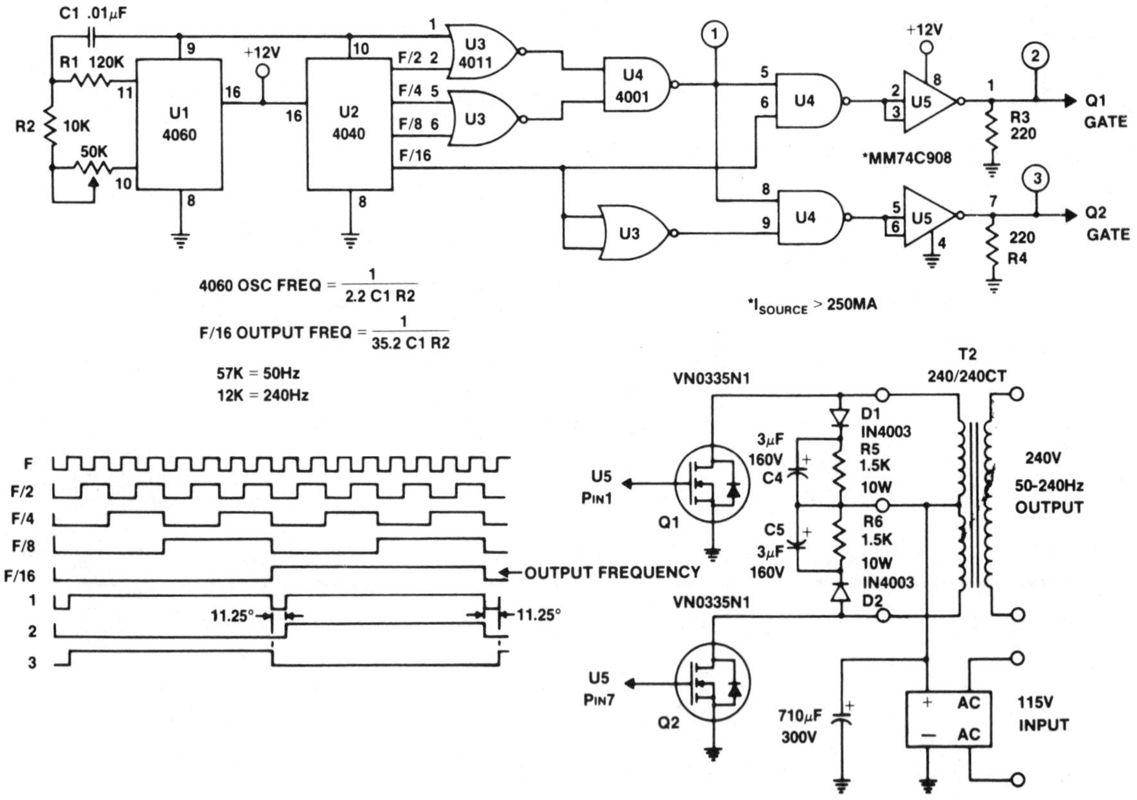

$$\text{4060 OSC FREQ} = \frac{1}{2.2\ C1\ R2}$$

$$\text{F/16 OUTPUT FREQ} = \frac{1}{35.2\ C1\ R2}$$

57K = 50Hz
12K = 240Hz

*$I_{SOURCE} > 250MA$

Fig. 10-30   (SU)

SIMPLIFIED TACTILE DISPLAY

Numerous product areas still await miniaturization. This application illustrates one such area, where logic level signals must control 120 separate 180 V piezo actuators, requiring 120 base resistors and 120 H.V. transistors on the board. A better way includes only 15 DMOS ARRAYS and 14 resistors networks, resulting in a PCB parts reduction of 331.

Besides the reduction in parts, an additional benefit is much better reliability. Bipolar devices often fail driving high voltage hi-capacitance loads such as piezo devices. DMOS is well suited with its current sharing ability and no second breakdown failures. The availability of the higher voltage AN01 is very desirable in this application, since piezoelectric components are very capable of generating high voltage when mechanically excited. An actuated brail-pin inadvertently pushed hard, will respond with a voltage spike across the electrically-off drive transistor with +180 V already across the display; possibly doubling V_{ceo} requirements. The 300 volt, AN0130NA DMOS ARRAY should prove very reliable in this circuit.

AN0130NA

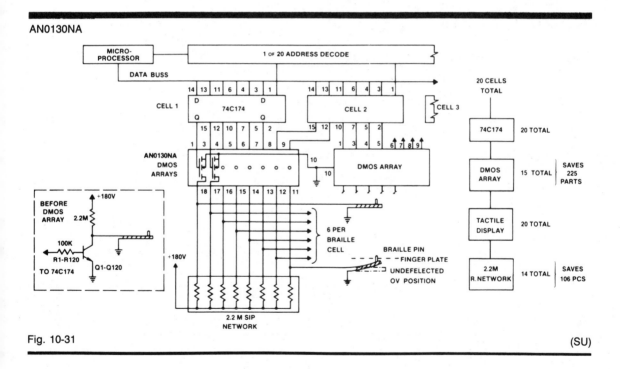

Fig. 10-31

(SU)

PWM TO AC POWER MODULE

This is a TTL controlled· pulse-width to voltage switching module. One suggested use inputs digital words representing audio signal levels, converts them to very fast transistion varying-width pulses that are inputed to the output module. The switching circuit performs the time to voltage conversion, with the summation of switched current over time in the LC networks. DMOS transistors provide a simple drive circuit and reliability.

Other applications include reversible motor speed controllers, switching dual-polarity system supplies, and high power waveform generators; all with direct digital control. A lower power version for synthesized voice output is also a possible application.

VN1306N3, VP1306N3, VN1110N1, VP1210N1

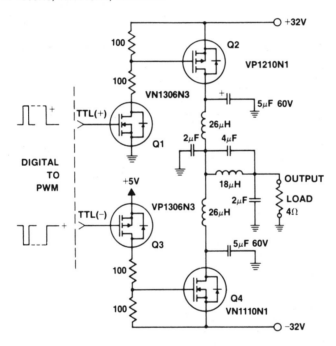

Fig. 10-32   (SU)

10 AMP FAST PULSE GENERATOR

This circuit takes advantage of DMOS fast rise time and minimum gate sustaining requirements. Output pulse risetime is as fast as 5 nanoseconds with VN1304 devices. An important feature of this circuit is that input pulses of shorter time duration than 100 nS, produce shorter duration output pulses with repetition rate determined by Rrep-Ct. Low standby power requirements make this application well-suited to battery operation.

VN1304N3, VP1204N2

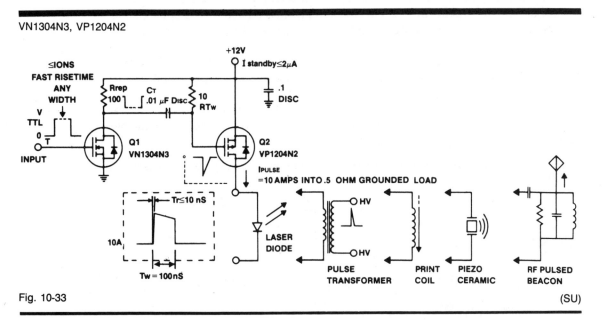

Fig. 10-33

(SU)

6-STATE DETECTOR/INDICATOR

This 6-state detector uses a single bicolor LED to detect and display six out of eight possible input conditions. The B and C inputs control the color displayed, while the A-input controls the flashing mode according to the table. The flash rate is controlled by R1 and C1 when the A-input is low. When all inputs are low the display will alternate red and green. The DMOS device, VN0104, controls the direction through the bicolor LED. R2 is selected at a lower value to provide more current for the green lamp in order to maintain equal luminescence.

There are several advantages of using the DMOS device in this application. The primary advantages are CMOS compatibility and cost effectiveness. The circuit is also power efficient making the circuit suitable for battery operation VDD can range anywhere from 3 to 18 V as long as R2 and R3 are adjusted accordingly. The circuit also has a minimal component count versus designs using linear devices, bipolar transistors, or TTL/LSTTL devices.

The DMOS device is also more suitable to sink the current necessary for the LED. Should the user desire to increase the brightness, this can be done by lowering the limit resistors R2 and R3.

VN0104N6

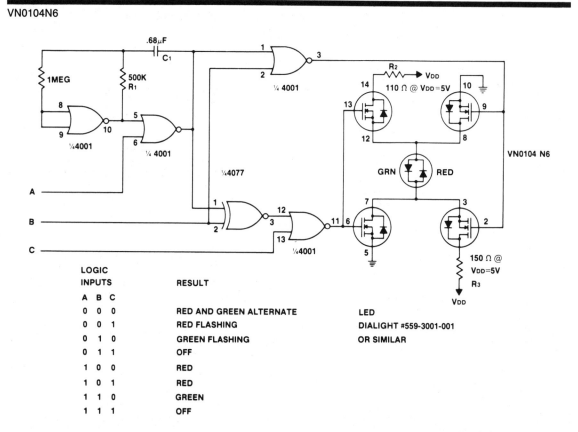

LOGIC INPUTS			RESULT	
A	B	C		
0	0	0	RED AND GREEN ALTERNATE	LED
0	0	1	RED FLASHING	DIALIGHT #559-3001-001
0	1	0	GREEN FLASHING	OR SIMILAR
0	1	1	OFF	
1	0	0	RED	
1	0	1	RED	
1	1	0	GREEN	
1	1	1	OFF	

Fig. 10-34

(SU)

FULL-SWING AC AMPLIFIER

Full ± 15 V output swing is accomplished without the increased complexity & inherrent instabilities of a common-source configuration by powering the OP AMP supply pins from the output signal. D1 and D2 limit the possible 30 volts swing from exceeding the HA 2525's maximum supply rating.

High frequency performance, which is OP AMP current and slew rate dependent, is improved with the optional bias network. Low frequency limitations are limited by the amount of charge in C1 & C2. The DMOS devices provide simple gate drive requirements and are highly reliable with complex loads. Applications include mobile sound systems and medium power portable radios.

VN1106N5, VP1206N5

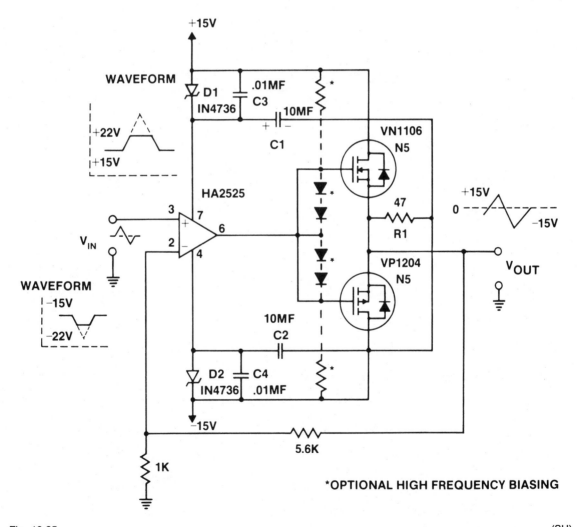

Fig. 10-35

MINI-SIZED LOW DROPOUT REGULATOR

Four components and a p-channel DMOS transistor, is all that is necessary to construct this precision variable regulator with up to 1 A output. Linear regulation benefits from DMOS's almost non-existent drive current (typically 1 nanoamp), and makes possible this circuit's stable lower power consumption (4.5 mw @ 9 V input). Low dropout voltage, and small size suggest uses in many types of portable equipment with battery power budgeting.

VP1204N5

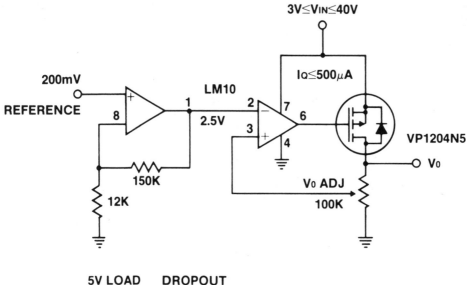

5V LOAD	DROPOUT
50Ω	0.1V
8Ω	0.4V
4Ω	0.8V

Fig. 10-36

SATELLITE TRANSMITTER DRIVER

Computers in industrial environments often need optical interface to eliminate electrical noise interference. This is mandatory on portable hand-held terminals now employed by many factories for machine control and inventory-materials update, as electrical cords are out of the question. Battery life is dictated mostly by the infrared array current drive circuitry that often requires 10 amp pulses, delivered with time-domain accuracy. This usually means additional standby current flowing through bipolar driver transistors and a shorter battery service cycle.

Typical standby power is 1 microwatt (4098B excluded). The output driver and DMOS transistor use power only during the output pulse period. Output delay time of the driver-DMOS combination is under 200 nanoseconds, fast enough for most time-domain data entry.

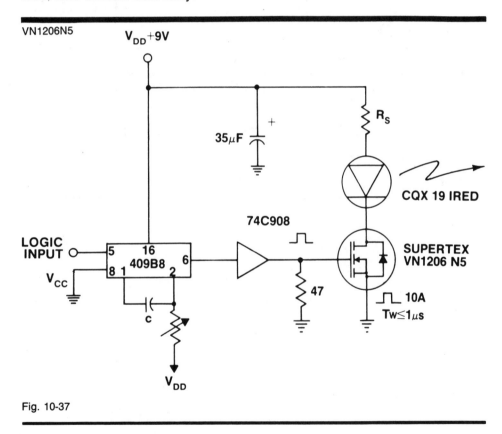

Fig. 10-37

AIR CLEANER HIGH VOLTAGE GENERATOR

This compact switching high voltage generator for an air cleaner is useful for generating power for a variety of input and output voltages. The 555 oscillator is powered by the same zener resistor divider supplying 10 volts directly to the gate of Q3. Q3's gate is quickly clamped off by Q1 and Q2 action, while the 555 output turns on Q4 more slowly allowing safe switching transistions.

VN0340N5, VP0340N5

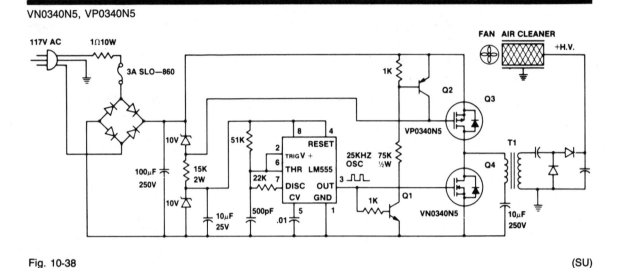

Fig. 10-38 (SU)

PRINT PEN SOLENOID DRIVER

Although this application uses two transistors for TTL to MOS level shift, the circuit is directly controllable from CMOS logic reducing parts needed. Low threshold TN0204 MOSFETs may be used for direct interface with TTL logic, when coil current requirements are lower.

VN1204N2, TN0204N3

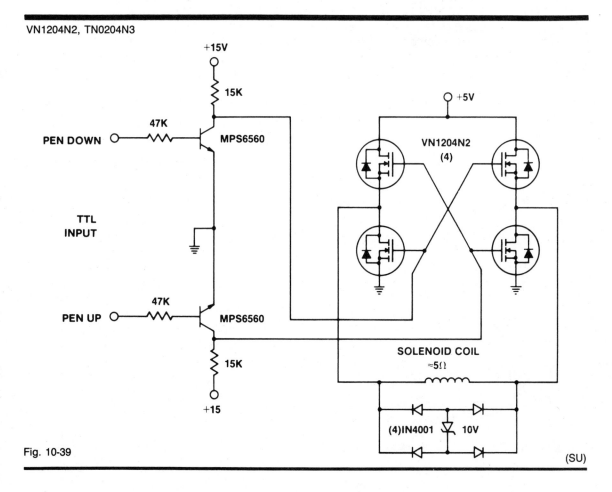

Fig. 10-39

(SU)

IMAGE TUBE TARGET-VOLTAGE CONTROLLER

In Vidicon cameras, the target voltage is often adjusted to obtain maximum resolution and sensitivity to the scene being viewed with the available light. The 7735 Vidicon tube shown requires 50 volts applied to the target at 1 lux incident illumination, on the Vidicon target and 10 volts for 10 lux illumination. This suggests that the controlling element for an adjustment circuit perform in a linear manner. As shown in the table, gate control voltage linearity is within ± 0.11 volts.

This circuit has the advantages of improved linearity, fewer circuit components, first order temperature correction, and less chance of failure than the circuits usually designed around bipolar transistors for target voltage control.

VN0545N3

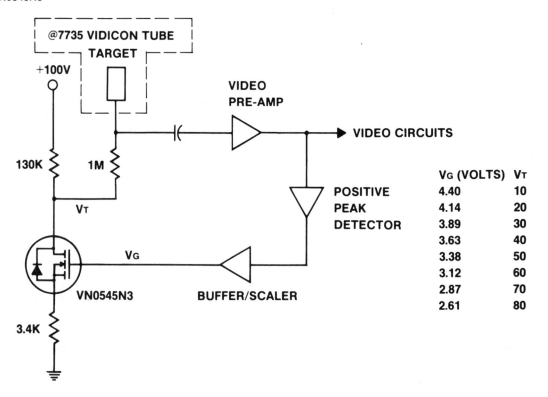

V_G (VOLTS)	V_T
4.40	10
4.14	20
3.89	30
3.63	40
3.38	50
3.12	60
2.87	70
2.61	80

Fig. 10-40

(SU)

PREKNOCK TRANSMITTER DRIVER CIRCUIT

The purpose of this circuit is to convert a TTL pulse at the input to a −100 volts output to a coaxial cable.

VP1310N3 provides an inversion, a level shift and voltage gain. R2/C2 determine the reset time and along with R3 determines all the timing parameters.

Applications for high voltage power pulses include acoustic rangefinders and ultrasonic cleaning equipment utilizing piezo electric transducers.

This specific design is intended for use in a Radar Magnetron driving scheme.

VP1310N3, VN1116N2

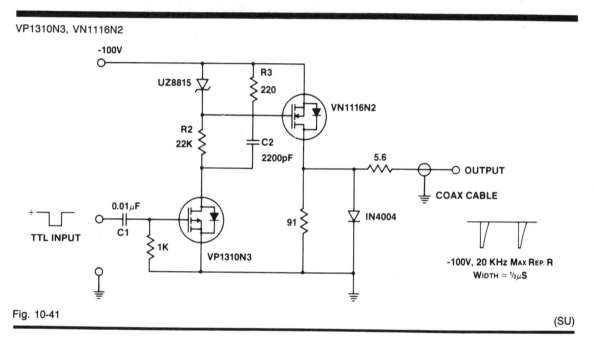

Fig. 10-41

(SU)

REVERSE BATTERY PROTECTION

The VN13 replaces a Schottky diode functioning as reverse battery protection. The reason why the VN13 body diode (specified at V_{SD} = 1 V @ 1.0 A, V_{GS} = 0) achieves a drop of 30 millivolts @ 10 mA, is that when V_{GS} = +6 V, V_{SD} decreases significantly, since it is shunted by R_{DS} (ON).

Assuming that the capacitor in the system is initially discharged, the integral source-drain diode provides the initial current to the system when the batteries are first attached. As the voltage across the load increases, the gate voltage also increases, turning on the DMOS transistor, reducing the drop across the transistor to a minimum.

If the batteries are placed in the circuit backwards, the transistor remains off, because of a 0 gate to source voltage. This protects the system from reverse current. Reversing the batteries while there is still positive charge on the system capacitor will cause the transistor to be on, until the capacitor is discharged.

VN1304N3

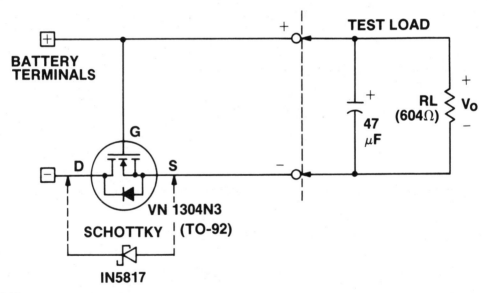

Fig. 10-42   (SU)

IMPROVING DATA CONVERTER
PERFORMANCE WITH INEXPENSIVE COMPONENTS

High performance at low cost is the primary feature of these reference-voltage circuit designs shown in Fig. 10-43. Although they have some drawbacks, circuit performance proves adequate for many applications.

An order-of-magnitude reduction in power-supply sensitivity is realized with this simple op-amp circuit of Fig. 10-44. Use of less expensive power supplies can also result in significant system cost savings.

Despite its simplicity, the resistor/diode compensation circuit in Fig. 10-45, improves converter temperature-drift performance nearly ten-fold. Adding an equal but opposite diode drift effectively cancels net converter drift.

Double the conversion speed with no loss in accuracy is what this simple diode-bridge clamping circuit in Fig. 10-46 offers. This circuit can also improve the performance of more sophisticated and expensive converters.

741   Improving data converter performance with inexpensive components

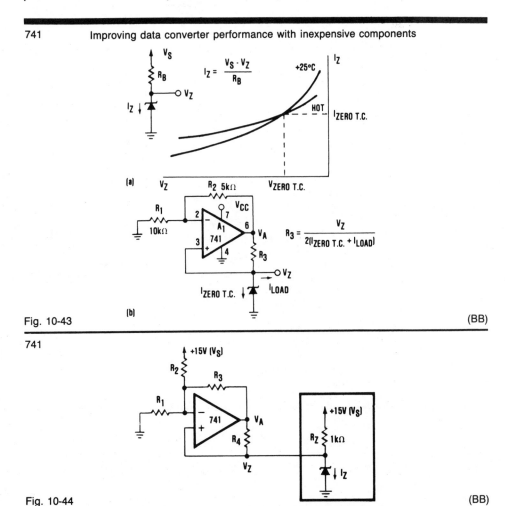

Fig. 10-43   (BB)

Fig. 10-44   (BB)

741

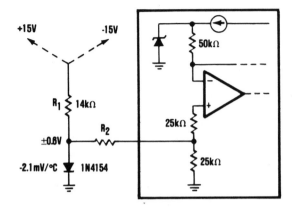

Fig. 10-45 (BB)

741

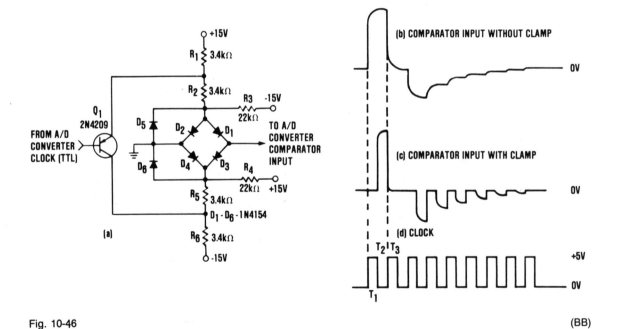

Fig. 10-46 (BB)

Appendix
Electronics Parts and
Components Suppliers

Albia Electronics
24 Albia Street
PO Box 1833
New Haven, CT 06508

All Electronics Corp.
905 S. Vermont Ave.
PO Box 20406
Los Angeles, CA 90006

Arrow Electronics, Inc.
25 Hub Drive
Melville, NY 11747

Digi-Key Corporation
PO Box 677
Thief River Falls, MN 56701

Fair Radio Sales Co., Inc.
1016 E. Eureka St.
PO Box 1105
Lima, OH 45802

Heathkit
Heath Company
Benton Harbor, MI 49022

H&R Corporation
401 E. Erie Ave.
Philadelphia, PA 19134

Jameco Electronics
1355 Shoreway Road
Belmont, CA 94002

Kelvin Electronics Inc.
PO Box 8
1900 New Highway
Farmingdale, NY 11735

Nuts & Volts
PO Box 1111
Placentia, CA 92670

Marlin P. Jones & Assoc.
PO Box 12685
Lake Park, FL 33403-0685

Mouser Electronics
PO Box 9003
Lakeside, CA 92040

R&D Electronics
1202H Pine Island Road
Cape Coral, FL 33909

Index

Index

Other Bestsellers of Related Interest

THE MASTER HANDBOOK OF IC CIRCUITS
—2nd Edition—Delton T. Horn

Crammed into this hefty handbook are 979 different circuits, using more than 200 popular ICs! Horn has revised and updated this bestseller to offer the latest IC designs and project ideas you can construct in your own home workshop at an amazingly low cost. Includes circuit design descriptions, applictions, pinout diagrams, schematics, and step-by-step instructions. 592 pages, 960 illustrations. Book No. 3185, $24.95 paperback, $34.95 hardcover

UNDERSTANDING ELECTRONICS—3rd Edition
—R.H. Warring, Edited by G. Randy Slone

Revised with state-of-the-art information on all the modern advances in electronics, this classic sourcebook is more completed than ever! You'll find thorough coverage of all the basics of electronics, and everything from ac and dc power to the developing new fields of photoelectronics and digital computing. This book offers all the information you need to begin designing and building your own circuits. 230 pages, 188 illustrations. Book No. 3044, $11.95 paperback, $18.95 hardcover

TROUBLESHOOTING AND REPAIRING ELECTRONIC CIRCUITS—2nd Edition—Robert L. Goodman

Here are easy-to-follow, step-by-step instructions for troubleshooting and repairing all major brands of the latest electronic equipment, with hundreds of block diagrams, specs, and schematics to help you do the job right the first time. You will find expert advice and techniques for working with both old and new circuitry, including tube-type, transistor, IC, microprocessor, and analog and digital logic circuits. 320 pages, 236 illustrations. Book No. 3258, $18.95 paperback, $27.95 hardcover

SCIENCE FAIR: Developing a Successful and Fun Project—Maxine Haren Iritz, Photographs by A. Frank Iritz

Here's all the step-by-step guidance parents and teachers need to help students complete prize-quality science fair projects! This book provides easy-to-follow advice on every step of science fair project preparation from choosing a topic and defining the problem to setting up and conducting the experiment, drawing conclusions, and setting up the fair display. 96 pages, 83 illustration. Book No. 2936, $13.95 paperback, $16.95 hardcover

OSCILLATORS SIMPLIFIED, with 61 Projects
—Delton T. Horn

Here's thorough coverage of oscillator signal generator circuits with numerous practical application projects. Pulling together information previously available only in bits and pieces from a variety of resources, Horn has organized this book according to the active devices around which the circuits are built. You'll also find extremely useful information on dedicated oscillator integrated circuits (ICs), an in-depth look at digital waveform synthesis, a clear description of the phase locked loop (PLL), and plenty of practical tips on troubleshooting signal general circuits. 238 pages, 180 illustrations. Book No. 2875, $14.95 paperback, $17.95 hardcover

ELEMENTARY ELECTRICITY AND ELECTRONICS
—Component by Component—Mannie Horowitz

This comprehensive overview of fundamental electronics principles uses specific components to illustrate explain each concept. You're led, step-by-step, through electronic components and their circuit applications. And because of its importance in today's technology, Horowitz has also included an introduction to digital electronics, complete with a description of number systems—decimal, binary, octal, hexadecimal. 350 pages, 231 illustrations. Book No. 2753, $18.95 paperback only

HOW TO USE SPECIAL-PURPOSE ICs—Delton T. Horn

A truly excellent overview of all the newest and most useful of the vast array of special purpose ICs available today, this sourcebook covers practical uses for circuits ranging from voltage regulators to CPUs . . . from telephone ICs to multiplexers and demultiplexers . . . from video ICs to stereo synthesizers . . . and much more! Easy-to-follow explanations are supported by drawings, diagrams, and schematics. 400 pages, 392 illustrations. Book No. 2625, $16.95 paperback only

AMPLIFIERS SIMPLIFIED, with 40 Projects
—Delton T. Horn

At last! A book on amplifiers that actually tells you what devices are used in which types of applications. Horn leads you through proper use of transistors, FETs, op-amps, and other DIP packages. He explains basic theory and covers problems commonly associated with amplifiers. Forty carefully selected projects provide hands-on experience in working with various types of amplifiers. 208 pages, 129 illus. Book No. 2885, $16.95 paperback only

101 SOUND, LIGHT AND POWER IC PROJECTS
—Charles Shoemaker

At last! Here's an IC project guide that doesn't stop with how and why ICs function . . . it goes one step further to give you hands-on experience in the interfacing of integrated circuits to solve real-world problems. Projects include sound control circuits such as alarms and intercoms; light control projects from photoflash slave to a monitor/alarm; power control units and much more. 384 pages, 135 illustrations. Book No. 2604, $16.95 paperback only

ENCYCLOPEDIA OF ELECTRONIC CIRCUITS
—**Volume 1**—Rudolf F. Graf

Here is every professional's dream treasury of analog and digital circuits—nearly 100 circuit categories . . . over 1,200 individual circuits designed for long-lasting applications potential. Adding even more to the value of this resource is the exhaustively thorough index which gives you instant access to exactly the circuits you need each and every time! 768 pages, 1,762 illustrations. Book No. 1938, $29.95 paperback, $60.00 hardcover

Prices Subject to Change Without Notice.

Look for These and Other TAB Books at Your Local Bookstore

To Order Call Toll Free 1-800-822-8158
(in PA, AK, and Canada call 717-794-2191)

or write to TAB BOOKS, Blue Ridge Summit, PA 17294-0840.

Title	Product No.	Quantity	Price

☐ Check or money order made payable to TAB BOOKS

Charge my ☐ VISA ☐ MasterCard ☐ American Express

Acct. No. _____ Exp. _____

Signature: _____

Name: _____

Address: _____

City: _____

State: _____ Zip: _____

Subtotal $ _____

Postage and Handling
($3.00 in U.S., $5.00 outside U.S.) $ _____

Add applicable state and local
sales tax $ _____

TOTAL $ _____

TAB BOOKS catalog free with purchase; otherwise send $1.00 in check or money order and receive $1.00 credit on your next purchase.

Orders outside U.S. must pay with international money order in U.S. dollars.

TAB Guarantee: If for any reason you are not satisfied with the book(s) you order, simply return it (them) within 15 days and receive a full refund.   BC